United Scholars Publications Series on Natural and Life Sciences

Advances in Natural and Life Sciences Volume: I

United Scholars Publications

United Scholars Publications, USA

www.unitedscholars.net

ISBN-13: 978-0692632222
ISBN-10: 0692632220

Preface

It is a major duty of researchers to keep up with the advances in life and natural sciences that occur almost daily. United Scholars Publications has recognized this duty and designed a book series that provide scientists, educators, and researches in this vital field with the cutting edge developments and to disseminate it globally.

This book, organized into seven chapters, features scientists from around the globe contributing diverse topics in mostly biological sciences. The book aims at highlighting the state of the art research and recent findings in modern biology and natural sciences and bridging theoretical research with applications. This collection will be of significant value to academic researchers, graduate students as well as practicing scientists working in this vibrant field.

In **Chapter One** the response of beam from retinal light stimulation and its gradual departure from the axial position of entry is evaluated. Moreover, retina's response to two departures; one, a gradual departure of the contrast in an interference pattern formed on the retina from unity; and the other, the departure of the spot-size ratio from perfect matching of unity are explored.

Chapter Two discusses new methods to improve the lactic acid production by subjecting lactic acid isolates to N_2 laser radiation.

Chapter Three investigates the possible beneficial effects of curcumin against nephrotoxicity induced by Carbon Tetrachloride in albino rats.

Chapter Four deals with the study of a mixture of the copolymer (PP/EPR) with calcium carbonates socal312, socal322v, and Winnofil spm from a thermal point of view. The chapter also highlights the effect of the addition of this component on the thermal properties of the above-mentioned copolymer.

Chapter Five discusses the RNA editing of miR-133, miR-208 and miR-151 which can open new avenues in the development of novel therapeutics. Loss of miR-133 in hypertrophy and heart failure due to an increased expression of miR-133 is also reported in this chapter.

Chapter Six reports on determining whether dexamethasone improves vascular dysfunction and morbidity in patients with this disorder whom are over 40 years of age.

Chapter Seven reports a study on the isolation and diagnosis of 30 isolates of the bacterium Shigella Spp which were collected from 180 stool specimens and rectal swabs of patients (children and adults). Samples were taken from subjects with watery and bloody diarrhea in AL- Miqdadiyah General Hospital and Al Zahra'a hospital in Iraq.

United Scholars Publications

January 2016

Table of Contents

CHAPTER ONE

COMPENSATIVE RESPONSE OF RETINAL CONES TO LOSS OF LUMINANCE EFFICIENCY UNDER OBLIQUE STIMULATION

Nachieketa K Sharma

Department of Physics, Institute of Technical Education & Research, Siksha 'O' Anusandhan University, Bhubaneswar-751030, Odisha, India

nachiketasharma@soauniversity.ac.in

ABSTRACT

A beam of light stimulates the retina strongly while entering through the center of the pupil. But gradual departure of the same beam from the axial position of entry on either side toward the edge elicits a poor visual response. This loss of luminance efficiency for oblique incidence is retina's response to the departure. Here, the response is evaluated by treating visibility loss as pupil apodization. In addition to this, retina's response to two more departures; one, a gradual departure of the contrast in an interference pattern formed on the retina from unity and the other, the departure of the spot-size ratio from perfect matching of unity are also explored. And it is found that the moment either the beam is gradually shifted toward the edge of the pupil or the intensity of one beam gradually departs from the intensity of the other beam or the spot-size ratio is away from unity the retina in each case invariably responds with a diminished visibility to suppress undesired light (intraocular stray light or light lost to minima of interference pattern or uncoupled light) in improving the image quality of photopic vision.

Keywords: Stiles-Crawford effect, coherent illumination, interference gratings, apodization, photoreceptor cone, waveguide, contrast, spot-size ratio.

INTRODUCTION

(A)Apodization: All the photoreceptors (both rods and cones) in the retina of a human eye share the same light coming from a single pupil. Still the light coming from different parts of the pupil stimulates the photoreceptors, especially the cones not in the same degree. Rather a beam passing near the edge of the pupil appears dimmer than a beam entering centrally. This is Stiles-Crawford effect of the first kind (SCE I) [1]. This diminution of visibility is also observed for a pupil that becomes less and less transparent from the centre toward the edge. This is apodization [2]. The SCE I has been widely treated as apodization [3,4] contrary to observation of the effect in photoreceptor cones only [5]. In the absence of aberrations the apodization approach gives identical results [6], provided one takes the apodization parameter twice that of the Stiles Crawford effect. Thus, a periodic target of specific transmission profile is chosen to compute the retinal response from the modulation transfer function [7,8] by modeling the SCE I as pupil apodization. Though targets of saw-tooth and triangular transmission profiles are used earlier to explore coherent response [9-12], a square wave grating is used to evaluate the retinal response in this work.

(B)Interference pattern on retina: The use of interference fringes for the exploration of the optics of the eye was initiated by Fergus Campbell and Daniel Green [13], though Arnulf and Dupuy were also credited with a similar work in French [14]. Two papers, one on the visibility of sine wave and square wave targets by DePalma and Lowry [15] and the other by Westheimer [16] on the modulation threshold for sine wave gratings on the retina gave the opportunity of synthesis to reach at the measurement of the quality of the eye's optics. Roland Gubisch polished the double pass setup for better measurements [17]. Then the contrast loss of the retina for oblique incidence was investigated by McMahon and MacLeod [18]. Westheimer also hinted at the phase distribution in Young's interference experiment [19]. Wijngaard worked on mode interference pattern [20]. With the advent of optical coherence tomography [21-23] and adaptive optics [24] retinal image analysis made a tremendous leap. In the present work, the two-wave interference is adopted to evaluate retinal response.

(C)Single-mode waveguide model of cone: The directionality involved in the cone-specific [5] nature of Stiles-Crawford effect of the first kind led Toraldo di Francia [25] to treat cones as radar waveguides which also exhibit directional sensitivity. The evidence of the retina being a microstructure with its cones working as waveguides of light was reported by Enoch [26-28]. Thus, SCE I was viewed as being created by a guiding of light into the cone photoreceptors [29]. Moreover, it has been experimentally observed that the optical fiber properties of a group of cones resemble those of an individual cone [30]. Incorporation of the diffraction properties of light into the waveguide model [6,31,32] then led, on one hand, to an analytical expression for the directionality coefficient (ρ_e) in agreement with the SCE I results [33,34] and on the other, to an exploration of physical rationale for the psychophysical SCE I [35].

METHODOLOGY

(A) Apodization: A vertical rectangular wave grating with average amplitude a, modulation b, period p and duty cycle α can be expressed as follows [36]:

$$A(x,y) = (a - b + 2\alpha b) + \frac{4b}{\pi}\sum_{n=1}^{\infty} \frac{sin(n\alpha\pi)}{n} cos(n\omega x) \quad (1)$$

Where $\omega = \frac{2\pi}{p}$ and x, y are the horizontal and vertical coordinates of space. Letting $(a - b + 2\alpha b) = C$ and $K_n = \frac{4b\, sin(n\pi\alpha)}{n\pi}$, taking the Fourier transform of Eq. (1) and using two dimensional Dirac delta function $\delta(p, r) = \iint_{-\infty}^{\infty} e^{-i(pq+rs)} dqds$ reduce Eq. (1) to

$$a(u, v) = C\ \delta(u, v) + \frac{1}{2}\sum_{n=1}^{\infty} K_n[\delta(u - n\omega, v) + \delta(u + n\omega, v)]$$
(2)

Given that the Fourier transform of the input to the pupil is $a(u, v)$, pupil exit function will be $f(u, v)a(u, v) = a'(u, v)$ for f being the amplitude transfer function which we approximate as follows [37,38]:

$$f(u,v) = \begin{cases} e^{\frac{-(u^2+v^2)}{\sigma^2}} & \text{if } u^2+v^2<K, \\ 0 & otherwise \end{cases}$$

Where K = 1 for diffraction-limited coherent imaging systems, and K = 2 for diffraction–limited incoherent imaging systems [37-40] for apodization parameter $\sigma = 3.086$ [41]

$$a'(u, v) = e^{\frac{-(u^2+v^2)}{\sigma^2}} \left[C\ \delta(u, v) + \frac{1}{2}\sum_{n=1}^{\infty} K_n[\delta(u - n\omega, v) + \delta(u + n\omega, v)]\right]$$
(3)

The image amplitude distribution at the exit pupil is obtained by the inverse Fourier transform of the spectrum $a'(u, v)$ giving rise to

$$A'(x, y) = C + \sum_{n=1}^{\infty} K_n \int_{-\infty}^{\infty}\int_{-\infty}^{\infty} \frac{[\delta(u-n\omega,v)+\delta(u+n\omega,v)]}{2} f(u, v)\, e^{i(ux+vy)} dudv$$

$$= C + \sum_{-\infty}^{\infty} K_n f(n\omega, 0) \cos(n\omega x) \quad (4)$$

Substituting back in for C and K_n yields

$$A'(x,y) = a - b + 2\alpha b + \frac{4b}{\pi}\sum_{n=1}^{\infty}\frac{sin(n\alpha\pi)}{n} f(n\omega,0)cos(n\omega x)$$

As $f(n\omega,0) = 0 for n\omega \geq 1$ the upper limit n' of n is such that $n\omega < 1$. So, the above equation becomes

$$A'(x,y) = a - b + 2\alpha b + \frac{4b}{\pi}\sum_{n=1}^{n'}\frac{sin(n\alpha\pi)}{n} f(n\omega,0)cos(n\omega x)$$

Finally, the image energy distribution will be given by the squared modulus of $A'(x,y)$ as

$$|A'(x,y)|^2 = \left[(a - b + 2\alpha b) + \frac{4b}{\pi}\sum_{n=1}^{n'}\frac{sin(n\alpha\pi)}{n} f(n\omega,0)cos(n\omega x)\right]^2$$

For a square wave grating of unit amplitude, as $a = b = \frac{1}{2}, \alpha = \frac{1}{2} \Rightarrow (a - b + 2\alpha b) = \frac{1}{2}$, the above equation becomes

$$\left[\frac{1}{2} + \frac{2}{\pi}\sum_{n=1}^{n'}\frac{sin(\frac{n\pi}{2})}{n} f(n\omega,0)cos(n\omega x)\right]^2$$

Where ωx is varied from 0 to π to compute maximum and minimum image energy distribution.

The modulation in the image (for unit contrast in the object) is

$$M = \frac{|A'(x,y)|^2{}_{max} - |A'(x,y)|^2{}_{min}}{|A'(x,y)|^2{}_{max} + |A'(x,y)|^2{}_{min}}$$
(5)

As the incoherent imaging is linear in image energy distribution the image energy distribution and the corresponding modulation equations for incoherent illumination are given by

$$I(x,y)_{incohernt} = a - b + 2\alpha b + \frac{4b}{\pi}\sum_{n=1}^{n'} \frac{sin(n\alpha\pi)}{n} f(n\omega,0)cos(n\omega x)$$

Where a is average intensity and b is modulation.

As $f(n\omega,0) = 0$for$n\omega \geq 2$ the upper limit n' of n is such that $n\omega < 2$ for incoherent illumination [38].

$$M_{inc.} = \frac{I_{max} - I_{min}}{I_{max} + I_{min}}$$
(6)

(B) Interference pattern on retina: The interference pattern formed on the retina with two coherent beams coming from the opposite edges of the pupil will have contrast(m) and relative phase(φ) as [41,42]

$$m = \frac{2\sqrt{I_1 I_2}}{I_1 + I_2} \text{and} \varphi = \tan^{-1}\left[-\left\{\frac{(m-1)-\sqrt{1-m^2}}{(m+1)+\sqrt{1-m^2}}\right\}(\tan kx)\right]$$
(7)

Where k denotes the magnitude of the incident wavevector component projected onto the retina and $\boldsymbol{x}$, is the retinal location. It is seen from Eq. (7) that the increase in contrast from $\boldsymbol{m} = \boldsymbol{0}$ to $\boldsymbol{m} = \boldsymbol{1}$ leads to the corresponding decrease in the underlying phase ($\boldsymbol{\varphi}$) (in radian) thereby establishing the intimacy between contrast and phase (or wavefront slope) computed for a fixed retinal location of $\boldsymbol{x} = \frac{\lambda}{8}$.As the wavefront slope is zero for axial entry and maximum for most peripheral entry, the pupil entry point of the beam also has a direct bearing upon the contrast through the following equation [42]

$$r_{eff} = -\left\{\frac{(m-1)-\sqrt{1-m^2}}{(m+1)+\sqrt{1-m^2}}\right\} r \quad (8)$$

Where 'r' is the pupil entrance point from the peak of the visibility.

In the absence of the interference pattern the traditional SCE I visibility (η) is given as [43]

$$\eta(r) = e^{-\rho_e r^2}$$

Where the directionality coefficient (SCE I) is ρ_e =0.115/mm^2.The replacement of r with r_{eff} gives the modified visibility (in presence of interference) as [44]

$$\eta(r, m) = e^{-0.115 r_{eff}{}^2} = e^{-0.115\left[-\left\{\frac{(m-1)-\sqrt{1-m^2}}{(m+1)+\sqrt{1-m^2}}\right\}r\right]^2} = e^{-0.115\left(\frac{1-m}{1+m}\right)r^2} \quad (9)$$

Eq. 9 shows the dependence of visibility on the contrast in addition to the pupil entry point.

(C) Single-mode waveguide model of cone: In an aberration free approximation a Gaussian beam whose intensity drops to 0.14 of its peak value at distance ω_o on either side of the peak value is focused on the retina to a spot- size of $\omega_r = \frac{\lambda f_{eye}}{\pi n_{eye}\omega_o}$ [33,45]. The parameters used are: focal length f_{eye}=2.22 cm, $\lambda = 6.328 \times 10^{-5}$ cm, a constant index of refraction n_{eye}=1.33 [46]. The focused beam couples its power at the retina to the fundamental mode as the V number is less than 2.405 for foveal cones with radius less than 2.218 λ [47]. As the incident beam couples light only to the fundamental mode represented by a Gaussian function of half-widthω_m (ω_m: waveguide mode spot-size) the fraction of power transmitted to the photoreceptor cone can be found as [47]

$$T(\theta) = \left[\frac{2\omega_r\omega_m}{\omega_r^2 + \omega_m^2}\right]^2 exp\left[\frac{-2\left(\pi n_{eye}\omega_r\omega_m\right)^2\theta^2}{\lambda^2(\omega_r^2 + \omega_m^2)}\right]$$
$$= \left[\frac{2\omega_r\omega_m}{\omega_r^2 + \omega_m^2}\right]^2 exp\left[\frac{-2\left(\pi n_{eye}\omega_r\omega_m\right)^2 r^2}{\lambda^2(\omega_r^2 + \omega_m^2)f_{eye}^2}\right]$$

$$\text{or} T(r) = \left[\frac{2\frac{\omega_r}{\omega_m}}{1+\left(\frac{\omega_r}{\omega_m}\right)^2}\right]^2 exp\left[-2\left(\frac{\pi n_{eye}}{\lambda f_{eye}}\right)^2\left(\frac{\omega_r^2}{1+\left(\frac{\omega_r}{\omega_m}\right)^2}\right)r^2\right]$$

Where $\theta = \frac{r}{f_{eye}}$, $\left(\frac{\omega_r}{\omega_m}\right)$ is the spot-size ratio and r is the distance from the on-axis value. With axial entry ($r = 0$),

$$T(\omega_r/\omega_m) = \left[\frac{2\frac{\omega_r}{\omega_m}}{1+\left(\frac{\omega_r}{\omega_m}\right)^2}\right]^2$$

(10)

So, the factor responsible for peripheral entry is

$exp\left[-2\left(\frac{\pi n_{eye}}{\lambda f_{eye}}\right)^2\left(\frac{\omega_r^2}{1+\left(\frac{\omega_r}{\omega_m}\right)^2}\right)r^2\right]$. Thus replacing ω_m with

$\omega_m = r_i\left(0.65 + \frac{1.619}{V^{\frac{3}{2}}} + \frac{2.879}{V^6}\right)$ [48] and comparing with $\eta = e^{-\rho_e r^2}$ [43], ρ_e is given as

$$\rho_e(r_i, \omega_r, V) = 2\left(\frac{\pi n_{eye}}{\lambda f_{eye}}\right)^2 \frac{\omega_r^2}{1+\left\{\frac{\omega_r}{r_i\left(0.65+\frac{1.619}{V^{3/2}}+\frac{2.879}{V^6}\right)}\right\}^2}$$

(11)

The power unable to be coupled to the photoreceptor cone due to departure of the spot-size ratio from unity can be found out by subtracting Eq. (10) from 1

$$1 - T(\omega_r/\omega_m) = 1 - \left[\frac{2\frac{\omega_r}{\omega_m}}{1+\left(\frac{\omega_r}{\omega_m}\right)^2}\right]^2 = \left(\frac{n^2-1}{n^2+1}\right)^2$$

(12)

Where $n = \frac{\omega_r}{\omega_m}$.

DISCUSSION

(A) Apodization: The image energy distribution in the images of periodic square wave targets, formed by a human eye apodized with SCE I under coherent illumination, is computed using Eq.(5) to find out the modulation in the images of periodic square wave targets plotted graphically in Fig. 1. Similarly, modulation is evaluated for the incoherent illumination using Eq. (6). The values are plotted in Fig.2.The Nyquist frequency, or the cut-off frequency determined by the geometry of the cone array is the highest spatial frequency that a normal human eye could resolve if it had perfect optics. So, the fineness of the retinal sampling array imposes a 75 Hz cut-off on square waves that can be delivered to the retina using incoherent light. And for coherent light this will be lower, i.e. 37.5 cycles/deg [38,49-55]. [Appendix I]

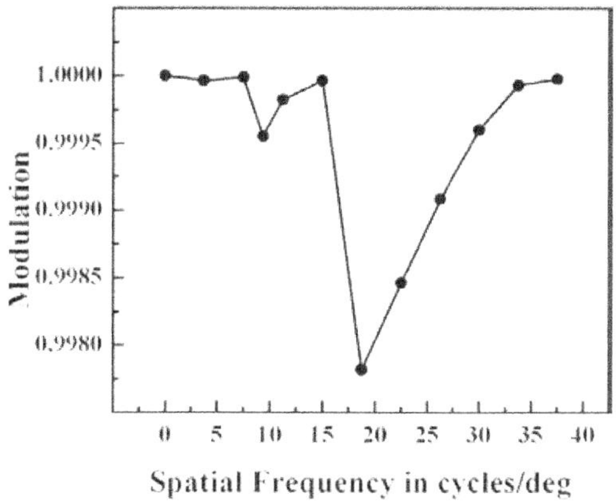

Fig.1. Variation of modulation with spatial frequency in cycles per degree of visual angle (ω) for coherent illumination for a human eye with SCE I modeled as pupil apodization.

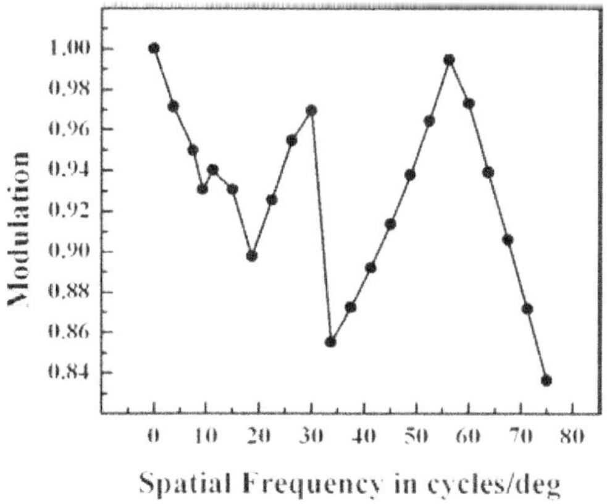

Fig.2. Variation of modulation with spatial frequency in cycles per degree of visual angle (ω) for incoherent illumination for a human eye with SCE I modeled as pupil apodization.

From Fig.1, it is evident that though the modulation initially decreases with the increase of the spatial frequency attaining a minimum contrast for a particular spatial frequency of 18 cycles/degree, finally increasing sharply with ω, the contrast changes only by 0.2 % throughout for coherent entering light. But from Fig.2, it is seen that the modulation registers a variation of as high as 20 % with respect to the spatial frequency for incoherent illumination. This is 100 times more compared to that for the coherent case. Again in Fig.3, the modulations for both coherent and incoherent illumination are compared for a human eye apodized with the Stiles Crawford effect of the first kind.

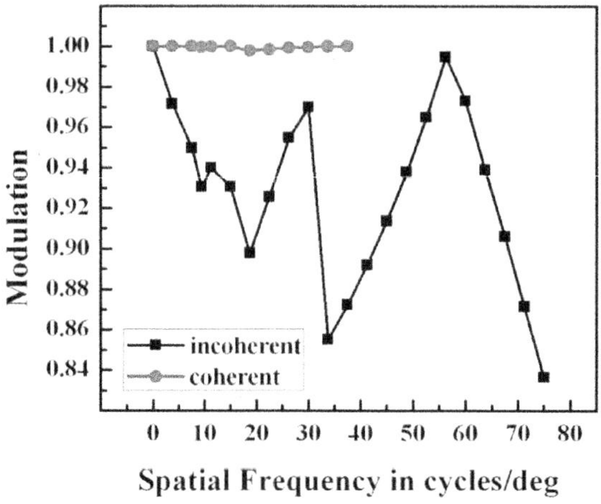

Fig.3. Variation of modulation with spatial frequency in cycles per degree of visual angle (ω) for both coherent and incoherent illumination for a human eye with SCE I modeled as pupil apodization.

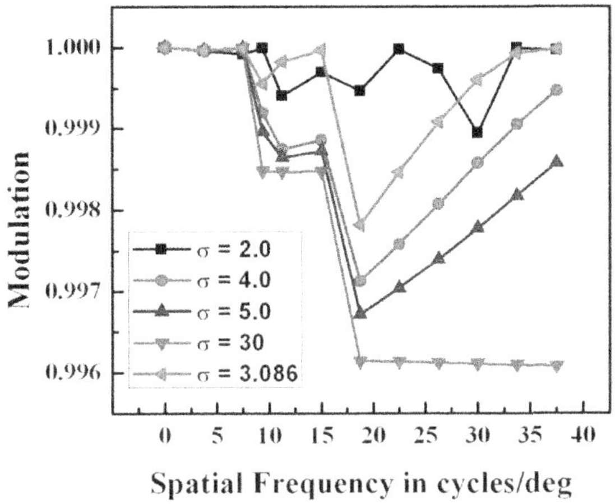

Fig.4. Variation of modulation with spatial frequency in cycles per degree of visual angle (ω) for different apodization parameters under coherent illumination.

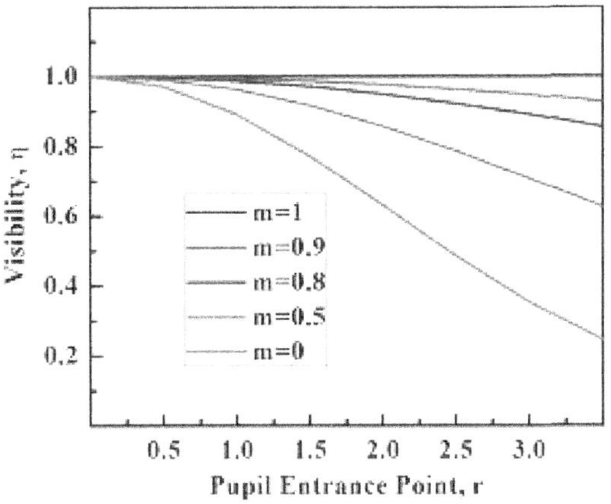

Fig.5. Variation of visibility with pupil entry point.

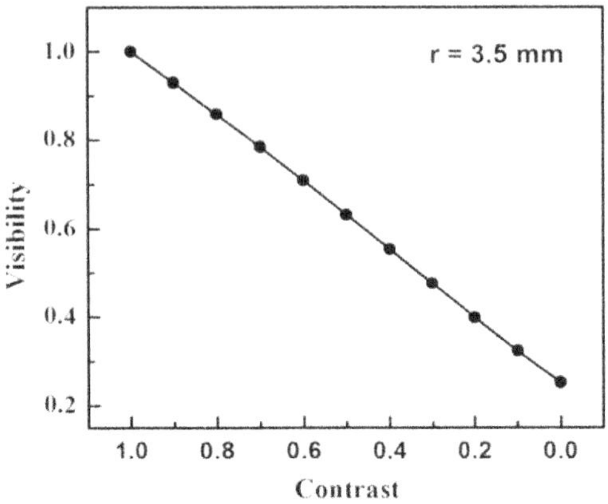

Fig. 6.Variation of visibility with contrast.

And it is found that for coherent illumination the retinal image energy distribution somehow remains insensitive to the SCE I apodization. Strikingly, with coherent illumination, the Stiles-Crawford effect is absent [3,9,56]. But the Stiles-Crawford effect operates in response to incoherent illumination to accentuate low vs. high spatial frequencies. In Fig.4, the modulation is plotted against spatial frequency for various values of the apodization parameter and it is noticed again that the contrast changes only by 0.4 % over the entire spatial frequency range for coherent illumination.

(B) Interference

pattern on the retina: From Eq. 9, $\eta(r,m) = e^{-0.115\left(\frac{1-m}{1+m}\right)r^2}$, it is evident that the visibility depends on pupil entry point (r) and contrast of the interference pattern (m) formed on the retina.

When visibility, η, is plotted versus the distance of the beam from the center of the pupil (r) for different values of the contrast (from m=1 to $m = 0$ through $m =$ 0.9, 0.8 and 0.5) three interesting observations are noted. First, with unit contrast in the interference pattern, the visibility does not depend at all on how the beams enter through the pupil. Secondly, with the disappearance of the interference pattern from the retina, the visibility follows the traditional SCE I. And finally, with the gradual decrease of the contrast from unity to zero, the visibility does not decrease that quickly with the increase of r compared to the traditional SCE I (fig. 5).When the contrast is slowly regulated from $m = 1$ to $m = 0$ (fig. 6) keeping the pupil location fixed at$r = 3.5$ mm, the visibility responds accordingly by decreasing to almost 0.24 of its maximum value.

(C) Single-mode waveguide model of cone: The uncoupled power 1-T (Eq. 12) is plotted against spot-size ratio in Fig. 7. For perfect matching, i.e. $\frac{\omega_r}{\omega_m} = 1$ the uncoupled power is zero or there is no visibility loss. But when the spot-size ratio departs from the perfect match either decreasing by a factor of half or increasing by a factor of two, the uncoupled power responds by increasing to 36 % of its peak value. Similarly, a mismatch of one- fourth or four times, i.e. when $\omega_m = \frac{\omega_r}{4}$or $\omega_m = 4\omega_r$, the uncoupled power becomes 78 % of the maximum coupled efficiency for perfect match. This is beautifully depicted in Fig. 7 and verified experimentally [33,34].

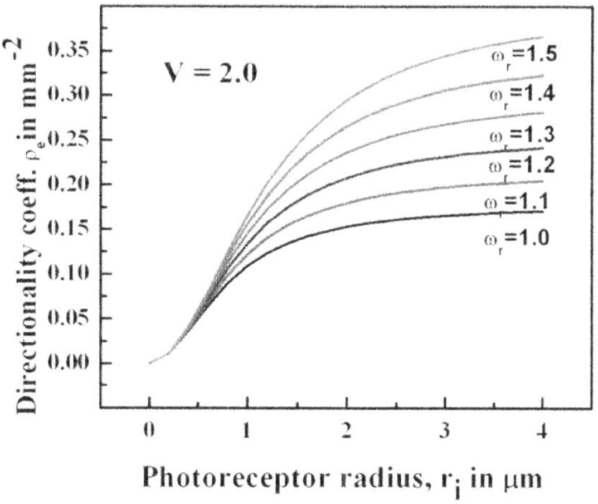

Fig.7. Response of power uncoupled to the fundamental waveguide mode to spot-size ratio. At perfect matching condition, i.e., $\omega_m = \omega_r$ (waveguide mode spot size equals the incident spot size) the uncoupled power is zero.

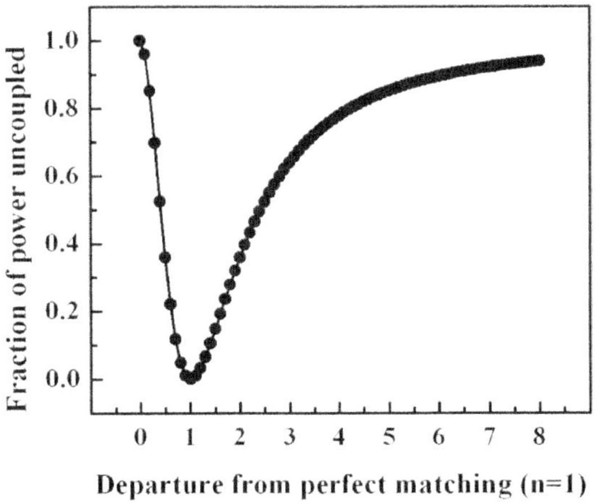

Fig.8. Variation of directional coefficient to the photoreceptor radius for different incident spot size. For a fixed radius, directionality increases with increase of ω_r indicating a strong directional coupling.

The variation of directionality coefficient ρ_e (in mm^{-2}) (Eq. 11) against photoreceptor radius r_i (in μm) is plotted in Fig. 8 for various incident spot-size ($\omega_r = 1.0, 1.1, 1.2, 1.3, 1.4$ & $1.5\mu m$) at fixed waveguide parameter of V = 2.0 for the fundamental mode. For f_{eye}=2.22cm, n_{eye}=1.33, $r_i = 1.2\mu m, \lambda = 6.328 \times 10^5 cm, \omega_r = 1.0\ \mu m, V = 2.0, \rho_e = 0.1236\ /mm^2$ or $\rho_{10} = \frac{0.053}{mm^2}$. This agrees with the observed SCE I directionality parameter [43] of $\rho_{10} = 0.053 \pm 0.012 mm^{-2}$.When the beam is well matched to the fundamental mode of the photoreceptor, the directionality parameter, ρ_e can be obtained as $\rho_c = \left(\frac{\pi n_{eye}\omega_r}{\lambda f_{eye}}\right)^2$. As revealed from Fig. 8, the directionality is highly dependent on the light coupling mechanism of the incident beam to the photoreceptor. Particularly in this case it is manifested in the relation between the spot-size of the incident beam and the radius of the photoreceptor as confirmed by the use of scanning laser ophthalmoscope [33, 34].

CONCLUSION

(A) Apodization: It is noticed that the modulation in the retinal image does not show any modification in the entire range of the spatial frequency for coherent illumination when the Stiles Crawford effect of the first kind is employed as pupil apodization for a target of square transmission profile. But when the entering light loses its spatial coherence the modulation undergoes modification. This suggests that the retinal image energy distribution depends on coherence of the entering beams and with the departure from perfect coherence the visibility diminishes accordingly. The same apodization that has been hypothesized to account for the SCE I, shows a one-hundred fold visibility changes with departure from coherent to incoherent illumination indicating retina's response to departure from coherence.

(B) Interference pattern on the retina: With the pattern no more present on the retina, its stimulation takes the traditional SCE route, but with a pattern of unit contrast the SCE I becomes totally irrelevant for the retina. Somehow, the formation of an interference pattern of unit contrast on the retina obliterates the Stiles-Crawford effect [3,42,56].

For intermediate situations ($m = 1$to $m = 0$) the visibility diminution with increasing pupil entrance point, characteristic of traditional SCE I, gets an opposing boost from the interference pattern on the retina in commensurate with its contrast. Thus, we see that with departure from unit contrast the visibility again diminishes.

(C) Single-mode waveguide model of cone: The spot-size of the incident beam (ω_r) is related to the radius of the photoreceptor (r_i) indicating a strong directional coupling of the individual cone photoreceptors to the focused incident beam. Thus, for a focused beam of light, the directionality turns out to be highly dependent on the spot-size ratio (ratio of the spot-size of the incident light and the size of the mode propagated along the photoreceptor). When the incident spot-size is identical to the fundamental waveguide mode spot-size (or the spot-size ratio is unity) the theoretical power, not able to be coupled to the cone, becomes zero. But with gradual departure of the spot-size ratio from this unit value for perfect matching (axial entry) there is loss of visibility for oblique incidence (peripheral entry).

Thus, it can be summarised that there is unmistakable visibility loss for departures either from perfect coherence or unit contrast and unit spot-size ratio. And the retina always compensates this loss for oblique incidence (various departures) through the Stiles Crawford effect of the first kind to suppress undesired light (intraocular stray light or light lost to minima of interference pattern or uncoupled light) in improving the image quality of photopic vision.

References

[1] W. S. Stiles and B. H. Crawford, "The luminous efficiency of rays entering the pupil at different Points," Proceedings of the Royal Society B, vol. 112, no. 778, pp. 428- 450, 1933.

[2] P. Jacquinot, P. Bougron and B. Dossier, "Calcul et realisation des distributions d'amplitude pupillaire, permettant a suppression des frangeslateraledans les figures de diffraction," in La Theorie des Images Optiques, P. Fleury, A. Marechal and C. Anglade, Eds. Paris: Editions de la Revue d'Optique 1949, pp. 183-193.

[3] G. Westheimer, "Retinal light distributions, the Stiles-Crawford effect and apodization," Journal of the Optical Society of America A, vol. 30, no. 7, pp. 1417-1421, July 2013.
[4] H. Metcalf, "Stiles-Crawford apodization," Journal of the Optical Society of America, vol. 55, no. 1, pp.72-74, Jan. 1965.
[5] F. Flamant, and W. S. Stiles, "The directional and spectral sensitivities of retinal rods to adapting fields of different wavelengths," Journal of Physiology, vol. 107, no. 2, pp. 187-202, Feb.1948.
[6] B. Vohnsen, "Photoreceptor waveguides and effective retinal image quality," Journal of the Optical Society of America A, vol. 24, no. 3, pp. 597-607, Mar. 2007.
[7] G. Westheimer, "Specifying and controlling the optical image on the retina, Progress in Retinal and Eye Research, vol. 25, no. 1, pp. 19-42, Jan. 2006.
[8] J. P. Carroll, "Apodization model of the Stiles Crawford effect," Journal of the Optical Society of America A, vol. 70, no.9, pp. 1155-1156, Sept. 1980.
[9] Nachieketa K Sharma, "Coherent Response of a Human Eye: An Apodization Model of Stiles Crawford Effect of the First Kind," Indian Journal of Science and Technology, vol. 8, no. 10, pp. 906–912, May 2015.
[10] Nachieketa K. Sharma, Kampal Mishra, Sushanta K. Kamilla, and Jatadhari K. Sharma, "Using Pupil Apodization as an Optical Technique to Obtain Retina Light Distributions in Presence of Stiles-Crawford Effect of the First Kind," International Journal of Microwave and Optical Technology, vol. 9, no. 3, pp.259-266, May 2014.
[11] Nachieketa K Sharma, "Modelling Stiles Crawford effect of the first kind as apodization," Orissa Journal of Physics, vol.22, no.1, pp. 89-98, Feb. 2015.
[12] Nachieketa K. Sharma, Kampal Mishra, Sushanta K. Kamilla, and Jatadhari K. Sharma, "Performance of a human eye in coherent illumination," Recent Advances

in Photonics (WRAP), 2013 Workshop on, IEEE Xplore, pp.1,2,17-18, Dec. 2013 doi: 10.1109/WRAP. 2013.6917630

[13] F. W. Campbell and D. G. Green, "Optical and retinal factors affecting visual resolution," Journal of Physiology, vol.181, no.3, pp. 576-593, Mar.1965.

[14] A. Arnulf and O. Dupuy, "La transmission des contrastes par le systemeoptique de l'oeil et les seuils des contrastesretiniens. C. R. Acad. Sci., Paris, vol. 250, pp. 2757-2759, 1960.

[15] J. J. DePalma and E. M. Lowry, "Sine-Wave Response of the Visual System. II. Sine-Wave and Square-Wave Contrast Sensitivity," Journal of the Optical Society of America, vol. 52, no. 3, pp. 328–335, Mar.1962.

[16] G. Westheimer, "Modulation thresholds for sinusoidal light distribution on the retina," Journal of Physiology, vol. 152, pp. 67-74, 1960.

[17] R. W. Gubisch, "Optical performance of the human eye," Journal of the Optical Society of America, vol. 57, no. 3, pp. 407-415, Mar.1967.

[18] M. J. McMahon and D. I. A. MacLeod, "Retinal contrast losses and visual resolution with obliquely incident light," Journal of the Optical Society of America A, vol. 18, no. 11, pp. 2692-2703, Nov. 2001.

[19] G. Westheimer, 'Phase distribution in Young's Interference Experiment,' Journal of the Optical Society of America, vol. 50, no. 12, pp. 1338, Dec. 1959.

[20] W. Wijngaard, "Mode interference patterns in retinal receptor outer segment," Vision Research, vol. 14, no. 9, pp. 889 – 893 (1974).

[21] Adolf F. Fercher, "Optical coherence tomography," Journal of Biomedical Optics, vol. 1, no. 2, pp. 157–173, Feb. 1996.

[22] David Huang, Eric A Swanson, C. P. Lin, J. S. Schuman, W. G. Stinson, W. Chang, M. R. Hee, T. Flotte, K. Gregory, C. A. Puliafito, and James G. Fujimoto,

"Optical coherence tomography," Science, New Series, vol. 254, no. 5035, pp. 1178-1181,1991.

[23] Niall Patton, Tariq M. Aslam, Thomas McGillivray, Ian J. Deary, Baljean Dhillon, Robert H. Eikelboom, ,KanagasingamYogesan, Ian J. Constable, "Retinal image analysis: Concepts, applications and potential," Progress in Retinal and Eye Research, vol. 25, no. 1, pp. 99–127, Jan. 2006.

[24] Austin Roorda, "Adaptive optics for studying visual function: A comprehensive review," Journal of Vision, vol. 11, no. 6, pp. 1–21, June 2011.

[25] Toraldo di Francia, G., "Retina cones as dielectric antennas," Journal of the Optical Society of America, vol. 39, no. 4, pp. 324, April 1949.

[26] J. M. Enoch and G. A. Fry, "Characteristics of a model retinal receptor at microwave frequencies," Journal of the Optical Society of America, vol. 48, no.12, pp. 899-911, Dec. 1958.

[27] J. M. Enoch, "Optical properties of the retinal receptors," Journal of the Optical Society of America, vol. 53, no. 1, pp. 71-85, Jan. 1963.

[28] J. M. Enoch, "Waveguide modes: Are they present, and what their possible role in the visual mechanism is?" Journal of the Optical Society of America, vol. 50, no. 10, pp. 1025-1026, Oct. 1960.

[29] A. W. Snyder and C. Pask, "The Stiles-Crawford effect: explanation and consequences," Vision Research, vol. 13, pp. 1115-1137, 1973.

[30] C. A. Curcio, K. R. Sloan, R. E. Kalina and A. E. Hendrickson, "Human photoreceptor topography," Journal of Comparative Neurology, vol. 292, no.4, pp. 497-523, Feb.1990.

[31] B. Vohnsen., I. Iglesias, and P. Artal, "Guided light and diffraction model of human-eye photoreceptors,"Journal of the Optical Society of America A, vol. 22, no. 11, pp. 2318-28, Nov. 2005.

[32] G. Westheimer, "Directional sensitivity of the retina: 75 years of Stiles-Crawford effect," Proceedings of the Royal Society B, vol. 275, no.1653, pp. 2777-2786, Dec. 2008.
[33] D. Rativa and B. Vohnsen, "Analysis of individual cone-photoreceptor directionality using scanning laser ophthalmoscopy," Biomedical Optics Express, vol. 2, no. 6, pp. 1423-1431, June 2011.
[34] B. Vohnsen and D. Rativa, "Ultra small spot size scanning laser ophthalmoscopy," Biomedical Optics Express, vol. 2, no.6, pp. 1597-1609, June 2011.
[35] Nachieketa K. Sharma and VasudevanLakshminarayanan, "Retinal response to departure from perfect power coupling: implications for the Stiles–Crawford effect," Journal of Modern Optics, vol. 62, no15, pp. 1278-1282, Sept. 2015.
[36] R. Barakat and A. Houston, "Modulation of Square-Wave Objects in Incoherent Light. I," Journal of the Optical Society of America, vol. 53, no.12, pp. 1371-1376, Dec. 1963.
[37] Artal, P. "Incorporation of directional effects of the retina into computations of optical transfer functions of human eyes," Journal of the Optical Society of America A, vol. 6, no. 12, pp. 1941-1944, Dec.1989.
[38] Joseph W. Goodman, "Introduction to Fourier Optics," Chapter 6, 3rd edition, Viva Books, New Delhi, 2007.
[39] G. Westheimer, "Retinal Light Distribution for Circular Apertures in Maxwellian View," Journal of the Optical Society of America, vol. 49, no.1, pp. 41-44, Jan. 1959.
[40] A.vanMeeteren, "Calculations on the optical modulation function of the human eye for white light," OpticaActa, vol. 21, no.5, pp. 395-412, 1972.
[41] Nachieketa K. Sharma, Kampal Mishra, Sushanta K. Kamilla, and Jatadhari K. Sharma, "Using Bacteriorhodopsin Thin Film in Regulating Retinal Stimulation of a Human Eye", Advanced Science Letters, vol. 20, no.3-4, pp. 705-709, March/April 2014.

[42] S. Castillo and B. Vohnsen, "Exploring the SCE-I with coherent light and dual Maxwellian Sources," Applied Optics, vol. 52, no.1, A1-A8, Jan. 2013.

[43] R. A. Applegate and V. Lakshminarayanan, "Parametric representation of Stiles-Crawford functions: normal variation of peak location and directionality," Journal of the Optical Society of America A, vol. 10, no.7, pp. 1611-1623, July 1993.

[44] Nachieketa K. Sharma; Contrast-controlled retinal response. Proc. SPIE 9654, International Conference on Optics and Photonics 2015, 965428, June 2015, doi:10.1117/12.2181663.

[45] Nachieketa K. Sharma, Kampal Mishra, Sushanta K. Kamilla, and Jatadhari K. Sharma, "Modelling of a Retinal Cone Using Erbium/Ytterbium Doped Polymethylmethacrylate and Crown Glass", Advanced Science Letters, vol. 20, no. 3-4, pp. 788-791, March/April 2014.

[46] H. H. Emsley, Visual Optics, 5th ed.; Butterworths: London; 1972; vol. 1.

[47] A. Ghatak, and K. Thyagarajan, Introduction to Fiber Optics, Cambridge, U.K., pp.149-156, 1998.

[48] D. Marcuse, "Loss analysis of single- mode fiber splices," Bell System Technical Journal, vol. 56, no.5, pp. 703-18, 1977.

[49] Warren Smith, Modern Optical Engineering, 4th Ed., McGraw-Hill Professional, 2007.

[50] Available from arapaho.nsuok.edu/~salmonto/vs2_lectures/Lecture19.pdf

[51] Raymond A. Applegate, "Limits to Vision: Can We Do Better Than Nature?" Journal of Refractive Surgery, vol. 1, pp. 547-551, Sept. /Oct. 2000.

[52] Available from www.swift.ac.uk/about/files/vision.pdf

[53] Larry N. Thibos and Arthur Bradley, "Modelling off-axis vision II: the effect of spatial filtering and sampling

by retinal neurons," School of Optometry, Indiana University Bloomington, Indiana 47405

[54] Rafael Navarro and M. Angeles Losada, "Phase transfer and point-spread function of the human eye determined by a new asymmetric double-pass method," Journal of the Optical Society of America A, vol 12, no.11, pp. 2385-2392, Nov. 1995.

[55] G. D. Boreman, Modulation Transfer Function in Optical and Electro-Optical Systems, SPIE Press, Bellingham, WA, vol. TT52, July 2001.

[56] B. Vohnsen and D. Rativa "Absence of an integrated Stiles-Crawford function for coherent Light," Journal of Vision, vol. 11, no.1, pp. 1-10, January 2011.

Appendix I

i) The diffraction cut-off frequency ($f_{cut\text{-}off}$) is given by $f_{cut\text{-}off} = 1/\lambda f/\#$, where f/# is calculated as given below.

$$f/\# = \frac{f}{D} \Rightarrow f_{cut\text{-}off} = \frac{D}{\lambda f} \text{ (for incoherent illumination)}$$

$$\text{and } f_{cut\text{-}off} = \frac{D}{2\lambda f} \text{ (for coherent illumination)}$$

f is the focal length, and D is the diameter of the entrance pupil (effective aperture).

When the diameter of the entrance pupil used is taken as 2 mm (D = 2 mm)

$f_{cut\text{-}off} = \frac{2}{555 \times 22.2 \times 10^{-6}} = 162$ lp/mm=48 cycles/degree (for incoherent illumination)

and$f_{cut\text{-}off} = \frac{2}{2 \times 555 \times 22.2 \times 10^{-6}} = 81$ lp/mm=24 cycles/degree (for coherent illumination)

Similarly, taking the diameter of the entrance pupil used as 3 mm (D = 3 mm)

$f_{cut\text{-}off} = \frac{3}{555 \times 22.2 \times 10^{-6}} = 243$ lp/mm=73 cycles/degree (for incoherent illumination)

and$f_{cut\text{-}off} = \frac{3}{2 \times 555 \times 22.2 \times 10^{-6}} = 122$ lp/mm=37 cycles/degree (for coherent illumination)

For a photopic vision pupil diameter of about (2-3) mm is taken. The performance of the human eye is near diffraction – limited. The other parameters are λ =555 nm, f_{eye}=2.22 cm. [0.3 is multiplied for conversion from lp (line pairs)/mm to cycles/degree.]

ii) The highest spatial frequency that the foveal cones can correctly resolve is the frequency whose half-cycle width is equal to the width of one cone. Based on the Nyquist theorem and the dimension of the foveal cones, the theoretical limit to spatial resolution, which is set by the diameter of a foveal cone is computed as follows:

a) Foveal cones are about 2 μm in diameter.
b) One cycle requires two cones, so the width of one cycle is 2 x 2μm = 4 μm
c) One degree on the fovea is about 300 μm wide.
d) How many cycles (each 4 μm wide) will fit into one degree (300 μm wide)? 300/4 = 75

$$\frac{\frac{300\mu m}{deg}}{\frac{4\mu m}{cycle}} = 75\frac{cycles}{degree}$$

This is the Nyquist frequency, or the cut-off frequency determined by the geometry of the cone array. In other words, this is the highest spatial frequency that a normal human eye could resolve if it had perfect optics. So, the fineness of the retinal sampling array imposes a 75 Hz cut-off on square waves that can be delivered to the retina using incoherent light. And for coherent light this will be lower, i.e. 37.5 cycles/deg.

CHAPTER TWO

EFFECT OF N2 LASER FLUENCE ON LACTIC ACID PRODUCTION BY LACTOBACILLUS BULGARICUS

Z. N. Al-karakooly[1,*] and Q. A. Al-anbaky[1,2]

[1]*Department of Biology, College of Science, University of Diyala. Diyala, Iraq*
[2]*Department of Biology, College of Sciences, Arts and Letters, University of Arkansas at Little Rock, Little Rock, AR 72204*
znalkarakoo@ualr.edu

Abstract

In previous years, the fermentation method has become more appropriate for the production of lactic acid, where the demand for naturally producing lactic acid has increased. The aim of this work is to improve the lactic acid production by subjecting lactic acid isolates to N_2 laser radiation. Fifteen samples of fermented dairy products were collected from local markets and dairy product manufacturers in Baghdad. Twenty four isolates were identified as bacillus lactic acid bacteria. All isolates had been subjected to the biochemical tests (catalase production, gelatin liquefaction, growth in litmus milk, NH_3 from arginine, CO_2 production, growth in 10 and 45 ^{o}C, growth at 4% NaCl, growth at 0.1% of methylen blue and fermentation of nine carbon compounds.) Efficient isolates for the production of lactic acid were selected and all isolates were propagated in whey medium. The results showed that *Lactobacillus bulgaricus* produced the highest total titrable acid (TA%) as it reached 1.31%.After treatment with N_2 laser at a fluence of 0.947 J/cm^2, production of lactic acid by this isolate reached 22.38 mg/ml compared to the control. Therefore, N_2 laser irradiation could efficiently enhance the lacic acid production ability of *L. bulgaricus*.

Key words: Lactic acid bacteria, N_2 laser, lactic acid

INTRODUCTION

Lactic acid is a natural organic acid (Reddy, et. al, 2008), which has a wide range of applications in pharmaceutical, food, leather and textile industries. It is used as an acidulant, preservative in food, flavoring agent and as a substrate to produce some organic acids (Li, et. al, 2010). In previous years, the fermentation process has become more appropriate for the production of lactic acid where the demand for naturally producing lactic acid has increased (Abdel-Rahman, et. al, 2013; Kadam, et. al, 2006). Moreover, the processing costs for the fermentation processes may be lower than those for synthetic processes (Vodnar, et. al, 2010). Scheele was first who discovered lactic acid in yogurt in 1780.

In 1857 Pasteur studies proved the presence of microorganisms producing lactic acid in yogurt (Foster, et.al, 1958). Both *Streptococcus* and *lactobacillus* are related to faculatatively anaerobic taxa evolved as individual lines of descent about (1.5–2) billion years ago when the earth passed from an aerobic to an anaerobic environment (Teuber and Erko, 1988).The genus *Lactobacillus* is a heterogeneous group of lactic acid bacteria with important implications in food fermentation (Giraffa, et.al, 2010) This group of bacteria have been used for decades in food preservation such as starters for dairy products and fermented vegetables

Lactic acid can be produced via chemical synthesis and microbial fermentation. Approximately 90% of lactic acid was produced by lactic acid bacteria (LAB) via fermentation of sugars. Today, lactic acid production has increased due to demand for its utilization as a new substance for biodegradable polymers (PLA) and renewable plastics (Schepers, et. al, 2006; Nancib, et. al, 2001). Recently, the quantity of lactic acid obtained by biotechnological methods has considerably increased (Romani, et. al, , 2008; Bonomo, et. al, 2008).

According to the modern development in bacterial taxonomy, lactic acid bacteria comprise several genera; *Streptococcus, Lactobacillus, Lactococcus, Pedococcus, Leuonostoic, Enterococcus, Aerococcus*. According to their metabolisms and similarities in the physical and nutritional needs, these genera are grouped together as lactic acid bacteria. One of the common major properties of these genera is the production of lactic acid as a main or virtually sole end product of the fermentation of sugar (Lndquest, 1998; Teuber, 1999).

A nitrogen (N_2) laser is a gas laser that produces extremely short (a few ns) intense (100 kW typical) pulses of light in the near UV portion of the E/M spectrum (337.1 nm). Nitrogen pumped dye laser was used in fluorescence microscopy, which is used to examine immunofluorescence technique by the existence of fluorescence dye (Goldman and Glibert,1982).A nitrogen pulsed lasers used in a microbeam technique to perform microsurgery in the cell contents with accuracy. The laser contents with microscopy to produce microradirradion of laser can be used as microdevice for cutting fixed parts of chromosomes (Leitz et. al,1994). In 1997 Pradeep et. al used N_2 laser for the evaluation of the diagnostic potential of human breast cancer tissues through using autofluorescence spectroscopy. They found in their research that the total integrated fluorescence intensity can provide an excellent discrimination between cancerous and benign/ normal breast tissues. Due to the important role of *Lactobacillus spp*in production of lactic acid, this study aimed to investigate the effect of an N_2 laser on the production of lactic acid and by *Lactobacillus bulgaricus*.

This research describes the use of N_2 laser to induce LAB isolates to explore their potential in the production of lactic acid. For the first time present data showed that metabolic targeted therapy with N_2 laser may improve the outcome of production of lactic acid by *L. bulgaricus*. This treatment modality can be easily introduced into experimental practice regulation.

Materials and Methods

Samples Collection and Isolation of Lactic Acid Bacteria

Samples were collected (in sterilized plastic packages) from dairy factories and local markets in Baghdad. Lyophilized starters (*Lactobacillus bulgaricus*) samples were brought to the laboratory in their sealed packages. MRS liquid medium was inoculated by 1% of milk and yoghurt samples, and incubated at 37°C for 24hr. Pour plate method was used for the isolation of lactic acid bacteria, according to Harrigan and MacCance (1976). After incubation, colonies surrounded by zones were selected. Isolated colonies were separately streaked on MRS agar for purification and then incubated at 37 °C for 24hrs under anaerobic conditions.

Identification Tests of Lactic Acid Bacteria

Bacilli isolates of LAB were identified depending on cultural, microscopical and physiological characteristics (biochemical tests), which include ammonia production from arginin, growth at 4% NaCl, growth in 0.1% of methylen blue, growth in 10 and 45 °C, carbohydrate sources fermentation and CO_2 production from glucose.

Selection of Efficient Isolates for Lactic Acid Production

Equal percentages from the suspension of the isolates which had been previously grown in an MRS medium are poured into flasks containing 100 ml sterilized whey with pH= 6.0. All flasks were incubated at 37 °C. The amounts of acid production were estimated. The pH value was daily determined by a pH-meter to keep it equal to 6 during the fermentation period by adding calcium hydroxide whenever needed.

Irradiation Methods

Isolates Preparation

Ten mls of MRS broth was inculcated with (1%) culture (3.8- 4.6x10^9 cell/ml) of each (*Lactobacillus bulgaricus*) isolate and incubated at 37 °C for 24hr. After incubation, the fermentation medium was centrifuged at 3000 rpm for 15 min. at 10 °C. Cells pellets were washed twice with phosphate buffer saline (PBS) (pH=7), then mixed by vortex and suspended in Tris- HCl, pH= 6.0 (Shaba, 2000).

N_2 -Laser Device

Pulsed N_2 laser (type Molectron UV 24) was used in this study. Its wavelength was 337.1 nm with 1.5 mJ per pulse and 10 ns pulse duration.

Laser Treatment

One milliliter of each prepared isolate (as mentioned in bacterial preparation) was suspended in Tris–HCl. They were transferred to sterile ependroff tubes. Nitrogen laser was used to treat the transferred bacteria. All samples were treated at room temperature in a dark place. Samples were then exposed to laser irradiation for the following energy densities (0.947, 18.95, 22.77 and 28.42) J/cm^2. After treatment, samples were put on ice to avoid DNA replication (Bauer, et. al, 1998).

The laser fluence was calculated from the following steps:
1. Energy Density (ED): Energy per unit area, i.e., ED (in J /cm^2) = E / A, where E is a single photon energy and A is the area.
2. Fluence (F) = ED. N, where N is the number of pulses per second and N = n.t. Here (t) is the exposure time in seconds and n is the number of pulses. The ependroff tube was located on a fixed base, while a mirror was used to adjust the reflected laser beam to the required area and a lens was used with a focal length of 5 cm to concentrate the laser beam on the bottom of the ependroff tube.

Detection and Determination of Lactic Acid

The procedure of Barker and Summerson (1954) was used for the detection and determination of lactic acid. Portions of 10 ml of the fermentation medium were put in sterilized separate test tubes, then centrifuged by 4000 rpm, for 20 min. Supernatant solution was removed and set in other sterilized test tubes. Subsequently, two milliliters of each of the supernatant solution of samples were added to 8 ml of trichloro acidic acid (10%) in sterilized separate test tubes, mixed thoroughly by a vortex. Turbid suspensions were appearing in all samples due to the protein denaturation, and so they were centrifuged by 3000 rpm for 10 min to get rid of proteins. Supernatant solutions were also put in sterilized test tubes. Then, five standard solutions of lactic acid were prepared with different concentrations (2, 4, 6, 8, and 10) mg/l (Fig. 1). Next two milliliters of supernatant from the centrifuged fermentation medium of each sample were taken and put in sterilized separate test tubes. Another 2 ml of five standard solutions was taken separately and put in another test tube. One ml of 20% copper sulfate solution was added and diluted to 10 ml, then mixed thoroughly by a vortex. Moreover, one gram of $CaCO_3$ powder was added to each tube stopper. They were shaken vigorously until the solution was uniformly dispersed and were allowed to stand for half an hour, then centrifuged with 3000 rpm for10 min. After that one milliliter from each supernatant solution was taken and put into sterilized test tubes, 0.05-ml of 4%copper sulfate solution was added to each tube. Next, 6 ml of concentrated sulfuric acid was added. The sulfuric acid should be added drop by drop at first. The contents of the tube were mixed well during the addition of acid. All tubes were placed upright in a boiling water bath for 5min, then transferred to cold water at 10 ^{o}C. Ten milliliters of parahydroxydiphenylre agent was added drop by drop to each tube. All tubes were placed in beaker continuum water at 30 ^{o}C for half an hour. Finally, all tubes were placed in boiling water for exactly 90 seconds, then in cold water at 10 ^{o}C or lower. The colored solution was transferred to a suitable container then the optical density was determined by the spectrophotometer model 20 at 560nm.

Statistical analysis

The results are expressed as mean ± SE from three independent experiments repeated in triplicate samples. Student's t test with $P \leq 0.05$ was used while calculating the level of significance.

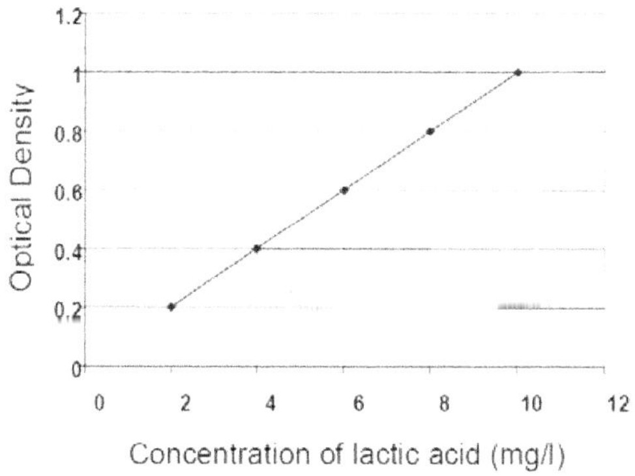

Fig. 1 Standard curve of lactic acid.

RESULTS AND DISCUSSION

Identification of LAB (Bacilli)

From the milk product samples locally collected, 24 isolates were primary detected as bacilli LAB according to the clear zone around their colonies on MRS agar containing 1% of $CaCO_3$. Colonies grown on the

MRS agar were circular, convex, slimy and white in color. Such characteristics are similar to the colonies formed by the bacilli genera of LAB. Microscopical examination of the LAB isolates showed that most of the isolates were bacilli cells arranged in long chains (7-10 cells), with presence of singles and pairs, they were G(+) non-spore forming. The MRS medium was very suitable for growth of LAB bacilli due to its contents of essential mineral salts that encourage the growth of this bacteria (De-Man, et. al, 1960; Tiwari, et. al, 1979).Biochemical tests were used for the differentiation between the different species of LAB(bacilli). Table1 (A and B) shows these tests.

Efficient Isolates for Lactic Acid Production

Figure (2) shows the production of lactic acid by the isolates grown in the fermentation medium (whey) expressed in percentages of total acidity. After 5 days of fermentation, some isolates which belonged to different species produced higher amounts of lactic acid than the other isolates of lactic acid bacteria. Depending on the most efficient isolates in acid production, one isolate has been selected for further study; it is *Lactobacillusbulgaricus* (Lb.13), which gave 1.36 of total acidity. This isolate was subjected to N_2 laser. Some investigators (Rodes and Fletcher, 1975; Casida, 1968; Teuber, 1993; Singh and Ranganathan, 1977) used *Strept.therophilus* to produce lactic acid by different production media (e.g. whey and molasses).

Table (1) Biochemical characteristics of the eleven samples of lactic acid bacteria isolates (LAB).

A- Fermentation of carbohydrate (CHO) sources

Isolates	Glucose	Lactose	Mannose	Mannitol	Arabinose	Xylose	Galectose	Maltose	Raffinose
Lb1	(1)+	(1)+	-	-	(w)	-	-	-	-
Lb2	(2)+	(1)+	-	-	-	(4)+	-	-	-
Lb3	(1)+	(1)+	(4)+	-	(4)+	-	(1)+	(1)+	(2)+
Lb4	(1)+	(1)+	(2)+	-	(2)+	-	(1)+	(1)+	(1)+
Lb5	(1)+	(1)+	(3)+w	-	(3)+	(4)+	-	(1)+	(1)+
Lb6	(1)+	(1)+	-	-	(4)+w	-	(1)+	-	-
Lb7	(1)+	(1)+	(2)+w	-	(3)+	(2)+	(1)+	(1)+	(2)+
Lb8	(1)+	(1)+	(1)+	(3)+	(4)+	-	(3)+	(1)+	(2)+
Lb9	(1)+	(1)+	(1)+	-	-	-	(1)+	(1)+	-
Lb10	(1)+	(1)+	(1)+	(1)+	(3)+	-	-	-	(2)+
Lb11	(1)+	(1)+	-	-	-	(3)+w	-	-	-
Lb12	(1)+	(1)+	-	-	-	-	-	-	-
Lb13	(1)+	(1)+	-	-	-	-	-	-	-
Lb14	(1)+	(1)+	-	-	-	-	-	-	-
Lb15	(1)+	(1)+	-	-	-	-	(1)+	-	-
Lb16	(1)+	(1)+	-	(4)+w	(4)+w	-	(1)+	(1)+	-
Lb17	(1)+	(1)+	(1)+	(2)+	-	(3)+	(1)+	(1)+	(2)+
Lb18	(1)+	(1)+	(1)+	(1)+	(2)+	-	(1)+	(1)+	(1)+
Lb19	(1)+	(1)+	(2)+w	(1)+	-	-	(1)+	(1)+	(1)+
Lb20	(1)+	(3)+w	(1)+	(1)+	(3)+w	-	(1)+	(1)+	(2)+
Lb21	(1)+	(1)+	(1)+	(1)+	(3)+w	-	(1)+	(1)+	(3)+
Lb22	(1)+	(1)+	(1)+	(1)+	(2)+w	-	(1)+	(1)+	(2)+
Lb23	(1)+	(1)+	(1)+	(1)+	-	(3)+	(1)+	(1)+	(2)+
Lb24	(1)+	(1)+	-	-	-			-	-

(+) positive reaction (-) negative reaction w= weak change with color () days of incubation

Inoculum Size and Treatments

Isolates Mutation

One selected lactics isolate (Lb.13) was irradiated with N_2 laser. Nitrogen (N_2) laser was used to treat (4.6×10^9) cells/ml of Lb.13. The following parameters of N_2 laser has been applied; (0.947, 18.95, 22.77 and 28.42) J/cm^2. Above inoculums size is usually regarded as suitable for similar kinds of studies. Singh and Ranganathan (1977) exposed 2×10^9 cells/ml of lactic cultures to gamma radiation (as a physical mutagen) and obtained high efficiency isolates in production of lactic acid (Hopwood,1970).

Production of Lactic Acid

Low power laser irradiation (LPLI) promotes proliferation of multiple cells, which is mainly through the activation of mitochondrial respiratory chain and the initiation of cellular signaling (Xuejuan and Da, 2009). So the fluence that leads to obtaining a maximum percentage of proliferation of the test bacteria (*Lb.bulgaricus*) was selected for production of lactic acid.

Twelve colonies belonging to *Lb.bulgaricus* growing on MRS agar were tested for production of lactic acid using Barker and Summerson's (1954) method shown in our procedure. The method depends on the standard curve, as in Fig (1). Despite that, most of the isolates that produced lactic acid were belonging to the parent bacteria. One isolate demonstrated high production of acid in comparison to the wild type isolate and other treatments. The isolate that was subjected to the fluences 28.42 J/cm^2 was producing 22.38 mg/ml of lactic acid. Fig. (3) illustrates the amount of lactic acid produced by the colonies of *Lb.bulgaricus*. Few studies are available in this field of investigation.

Mirganieva (1971) found that the production of lactic acid in whey medium increased by *Lb.bulgaricus* and its isolates after treatment with x- ray under a temperature of 46 ^{o}C, pH= 5.5- 6.0 with the absence of oxygen (Mirganieva,1971). While Al-Azzawi (1995) used ultraviolet light for improving the production of lactic acid in *Lb.bulgaricus* and *Strept.thermophilus* in whey medium and found that one of the mutated isolates was highly efficient in producing remarkable concentrations of lactic acid (Mahdi,1982).

B- Other biochemical characteristics.

Isolates	Catalase	Gelatin	NH3 from Argnine	Growth in litmus milk	Growth at 45 °C	Growth at10 °C	Growth in 4% NaCl	Growth in 0.1% methylene blue
Lb1	-	-	-	+	+	*	*	*
Lb2	-	-	-	+	+	*	*	*
Lb3	-	-	-	+	-	*	*	*
Lb4	-	-	-	+	-	*	*	*
Lb5	-	-	+	+	+	*	*	*
Lb6	-	-	-	+	+	*	*	*
Lb7	-	-	+	+	+	*	*	*
Lb8	-	-	-	+	+	*	*	*
Lb9	-	-	-	+	-	*	*	*
Lb10	-	-	-	+	-	*	*	*
Lb11	-	-	-	+	-	*	*	*
Lb12	-	-	-	+	+	*	*	*
Lb13	-	-	-	+	+	*	*	*
Lb14	-	-	-	+	+	*	*	*
Lb15	-	-	-	+	-	*	*	*
Lb16	-	-	-	+	+	*	*	*
Lb17	-	-	-	+	-	*	*	*
Lb18			-	+	-	*	*	*
Lb19	-	-	-	+	+	*	*	*
Lb20	-	-	-	+	-	*	*	*
Lb21	-	-	-	+	+	*	*	*
Lb22	-	-	-	+	-	*	*	*
Lb23	-	-	-	+	-	*	*	*
Lb24	-	-	-	+	+	*	*	*

(+) positive reaction (-) negative reaction (*) unable to grow

Mahdi (1982) mentioned that a greater amount of production of lactic acid was achieved when the pH value for whey medium was fixed at 6. Similar findings were obtained by (Marshall, 1970; Kempe et. al, 1950; Tiwari et. al, 1979). They found that a production of a greater amount of lactic acid is achieved at pH ranging between 5.5-6.0 and any increase or decrease from that range will lead to a decrease in the production of the acid amount (Kempe et. al, 1950; De-Man et. al, 1960; Marshall, 1970). For this purpose, Ca $(OH)_2$ was used in this study for controlling pH changes during fermentation.

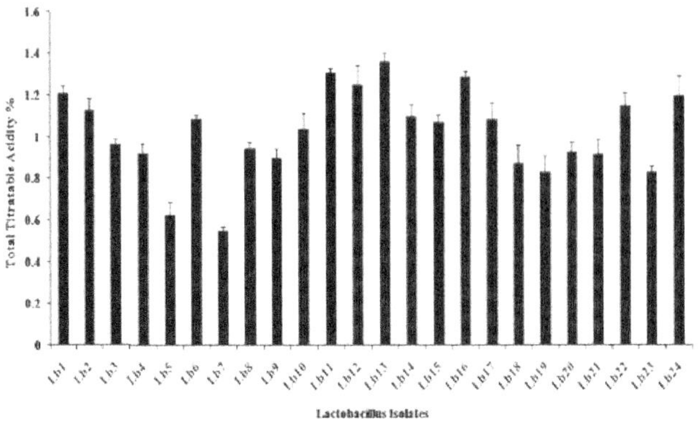

Fig.2. Total acidity of lactic acid production by *Lactobacillus Spp.* The values shown are mean ± SD from at least three independent experiments each performed in triplicates.

The effect of N_2 laser fluence on the concentration of lactic acid is illustrated in Fig. (3). The highest concentration of acid was obtained for the *Lb. bulgaricus* isolate at fluence of 0.947 J/cm^2. The effect of N_2 laser of 337.1 nm wavelength is most likely photochemical due to the absorption of laser light at this certain wavelength and certain energy densities by certain chromophores and lead to either biostimulation or bond breaking as follows: If the photon is absorbed by themolecule and its energy is less than that of the bond, the molecule should be affected. The excitation mechanisms lead to activate the molecule as:

$$(A\text{-}B) + h\upsilon = (A\text{-}B)^*$$

Where (A-B) is a molecule consists of two atoms, hυ is the photon energy and (A-B)* is the excited molecule. If the photon energy is equal or higher than the dissociation energy of the bond, it may be broken and produce energy as:

$$(A\text{-}B)^* + h\upsilon = A + B + EKIN$$

Table (3) illustrates the dissociation energy of the bonds, while Table (4) shows the photon energy for some lasers. The effect of laser radiation on the cells depends on its wavelength. This effect is either direct like when the wavelength is located in the UV region 190 to 280 nm where the chromophore or molecule that absorbs radiation is DNA, or it is indirect such as when the wavelength is located in the UVA region (320- 400 nm).

There is no effect on DNA at this wavelength of the laser beam because DNA doesn't absorb laser light in this wavelength, so the addition of specific chromophor to these wavelengths between (320-400) nm will induce the effect on DNA and is considered as exogenous chromophores (DeWith and Greulich,1995). The most probable chromophore which absorb the light of the N_2 laser (337.1 nm) represents the reduced form Nicotinamide Adenine Dinuclatide Phosphate NADP(H) (Fig. 5), where NADP (H) acts as a donor for H atoms or acceptor electrons in reductive biosynthesis reactions. It also enters into Krebs cycle and pentose phosphate pathways for producing Adenosine Tri-Phosphate (ATP) (Simcth, 1985). It is clear that the above description is coincidence with the conclusions of (Pukhoval and Salmon, 1995) which concluded that the NADP (H) of the macrophage cells absorbs N_2 laser light with 337.1 nm (Pukhoval and Salmon, 1995).

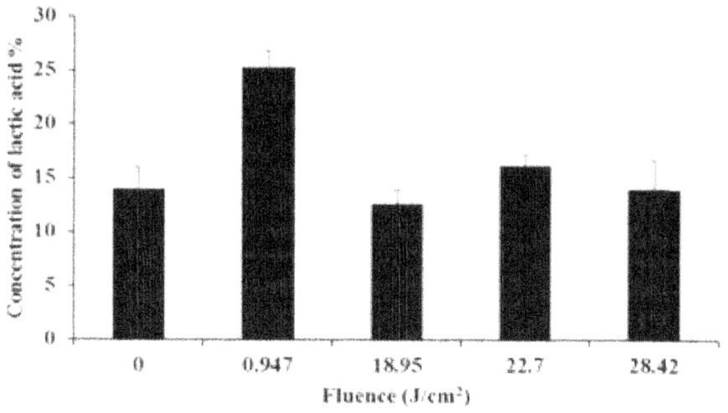

Fig.3. Fluence effect of N_2 laser treatment on the concentration of lactic acid that is produced by *Lb. bulgaricus* isolates. The values shown are a mean ± SD from at least three independent experiments each performed in triplicates.

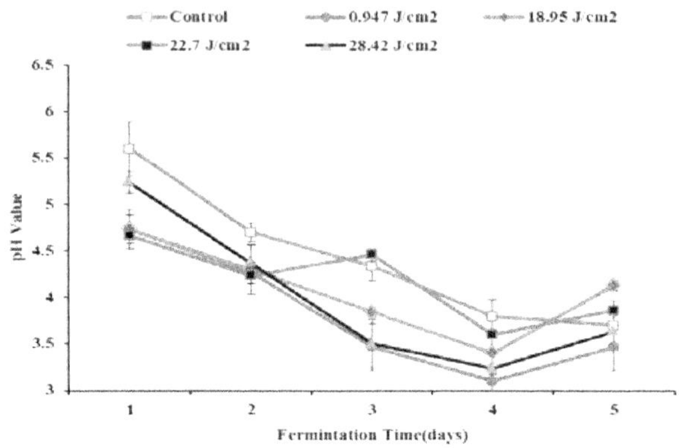

Fig 4: Change in pH values during propagating *Lactobacillus bulgaricus* in whey medium contains minerals slats at different fluences. The values shown are a mean ± SD from at least three independent experiments each performed in triplicates.

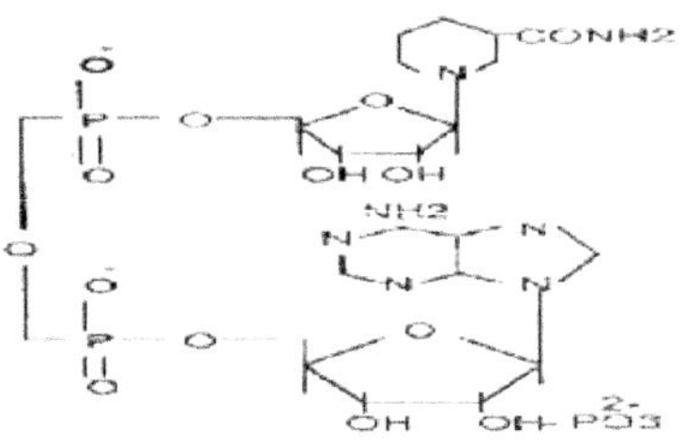

Fig. 5. Compound of Nicotinamide Adenine Dinucleotide Phosphate (Reduced form).

Table (4) illustrates that the photon energy of the N_2 laser is 3.6 eV, so the probable bonds that may be affected during the interaction are the bonds between (C-C, C-O, C-N) and as mentioned above either biostimulation or bond breaking occur according to the type of interaction.

Table 3: Energy dissociation of some bonds.

Type of bond	Dissociation Energy(eV)
C=O	7.1
C=C	6.4
O-H	4.8
N-H	4.1
C-O	3.6
C-C	3.6
C-N	3

Table 4: The wavelength and photon energy of some types of lasers.

Laser type	Wavelength nm	Photon energy eV
ArF	193	6.4
KrF	248, 268	5.0, 4.9
Nd:YLF 4W	263	4.7
XeCl	308	4
N_2	337.1	3.6
Argon Ion	488, 514	2.4
Nd:YLF 2W	526.5	2.3
He - Ne	633	2
Diode	800	1.6
Nd:YAG	1064	1.2
HO:YAG	2120	0.6
Er:YAG	2940	0.4

CONCLUSION

N_2 laser irradiation technology was employed to improve the lactic acid production ability from lactic acid bacteria. The optimal irradiation parameter was 0.947 J/cm^2, this was used to produce the maximum production of lactic acid. A positive strain (Lb13) could produce lactic acid under the optimal laser irradiation conditions. There are not enough studies on the mechanism of production of lactic acid by N_2 laser. So we need further studies to confirm these results. Also, N_2 laser irradiation could efficiently enhance the lactic acid production ability of *Lactobacillus bulgaricus*.

REFERENCES

[1] Abdel-Rahman, M.A., Y. Tashiro and K.Sonomoto (2013). Recent advances in lactic acid production by microbial fermentation processes. Biotechnology Advances. 31:877-902.

[2] Al-Azzawi, N. H. (1995) Isolation and Selection of Bacterial mutant producer of Lactic Acid. MSc. Thesis, College of Science, Al-Mustansriya University. 87 P.

[3] Barker, S. B. and Summerson, W. H. (1954) The colometric determination of lactic acid in biological materials. Journal of Biological Chemistery.138, 535.

[4] Bauer, E; Recknagel, R. D.; Fiedler, U.; Wollber, L.; Bock, C. and Greullich, K. O. (1998).The distrbution of the trial moment in single gel electrophoresis (comet assay) Obyes a chi-sequer (X2) not agaussiandistrbution. Mutation. Research. 398: 101-110

[5] Bonomo, M.G.A., T. Ricciardi, E. Zotta and G. Parente,(2008). Molecular and technological characterization of lactic acid bacteria from traditional fermented sausages of Basilicata region. Meat Sciences., 80: 1238-1248.

[6] Casida, L.E. J. (1968) Industrial microbiology. John-Wiley and Sons, N.Y.

[7] De-Man, J.C., Rogosa, M. and Sharpe, M. E. (1960) A medium for the cultivation of lactobacilli Journal AppliedBacterology, 23: 130-135.

[8] DeWith, A. and Greulich, K.O. (1995) Wavelength dependence of laser-induced DNA damage in Lymphocytes observed by single cell gel electrophoresis. Journal of PhotochemisteryPhotobiology. B-Bio. 99: 1- 6.

[9] Foster, E. M., Nelson, F. E., Speck, M. L., Dottsch, R. N. and Olsen, J. C. (1958) The microorganisms of milk and dairy products,in "Dairy Microb." MacMillian, London.

[10] Giraffa, G., N. Chanishvili and Y. Widyastuti, (2010). biotechnology, Research Microbiol., 161: 480-487.

[11] Goldman, O. J. C., and Glibert, P. M. (1982). Corporative rapid ammonium uptake by four marine phytoplankton species. *Limnology* and *Oceanography*. 27: 814-827.

[12] Harrigan, W. F. and MacCance, M. E. (1976) Laboratory Methods in Food and Dairy Microbiology. Acadm. Press, London.

[13] Hopwood, D. A. (1970) The Isolation of Mutants,in Methods in Microbiology Vol. 3

[14] Kadam,S.R., S.S.Patil, K.B.Bastawde, J.M.Khireand D.V.Gokhale, (2006). Strain improvement of Lacto-bacillus delbrueckii NCIM 2365 for lactic acid production. Process Biochemistery, 41: 120-126.

[15] Kempe, L. L., Halverson, H. O. and Pairt, E. L. (1950) Effect of continuously controlled pH on lactic acid fermentation yield and conversion.Industerial Engineering Chemistery, 42: 1852-1857.

[16] Leitz G, Weber G, Seeger S, andGreulich KO. (1994). The laser microbeam trap as an optical tool for living cells. Physiological chemistry and physics and medical NMR. 26(1):69-88.

[17] Li, Z., L.U. Han, Y. Ji, X. Wang and T. Tan, (2010). Fermentative production of lactic acid from hydrolysate of wheat bran by Lactobacillus rhamnosus., Biochemistery Engineering, 49: 138-142.

[18] Lndquest, J. (1998) General overview of LAB"Lactis, in laboratory manual for the food microbiology laboratory, John L.(Edt.) Wisconcen Madison.

[19] Mahdi, A. S. (1982) Role of microorganisms in production of lactic acid. M.Sc. Thesis, Agriculture College, University of Baghdad. 125 P.

[20] Marshall, K. R. (1970) Continuous production of lactic acid. Industrial of Dairy Congeress.

[21] Milganiveva, M. (1971) Effect of g - radiation on acid fermentation by thermophilic lactobacilli. Journal of Dairy Scienc, 39: 1611-1612.

[22] Mirganieva, M (1971). Effect of r-radiation on acid fermentation by thermophilus lactobacilli. J. Dairy Science, 39: 1611-1612

[23] Nancib, N., A. Nancib, A. BoudjelalC. Benslimane,. F. Blanchard, and J. Boudrant, (2001). The effect of supplementation by different nitrogen sources on the production of lactic acid from date juice by Lactobacillus casei subsp. Rhamnosus, Bioresour. Technol., 78: 149-153. whey by Lactobacillus bulgaricus. African Journal

[24] Pradeep K. G., Shovan K. M. andAbha U., (1997). Breast cancer diagnosis using N_2 laser excited autofluorescence spectroscopy Lasers in Surgery and Medicine. 21, 5: 417–422.

[25] Pukhoval, I. and Salmon, V. V. (1995) The effects of N_2 laser radiation on the Kinetics of the generation of the active forms of oxygen by the granulocytic-Macrophagal cells in a whole blood system, Radiatsionnaiabiologiia, radioecologiia, 35: 286- 291.

[26] Reddy, G., M. Altaf, B.J. Naveena, M. Venkateshwar and E. VijayKumar, (2008). Amylolytic bacterial lactic acid fermentation, Biotechnology. Advances, 26: 22-34.

[27] Rodes, A. and Fletcher, D. L. (1975) Principles of Industrial Microbiology, Pergamon Press. N.Y.

[28] Romani, A., R. Yanez, G. Garrote and J.L. Alonso, (2008). SSF production of lactic acid from Cellulosic biosludges. Bioresour. Technology, 99: 4247-4254.

[29] Schepers, A.W., J. Thibault and C.H. Lacroix, (2006). Continuous lactic acid production in whey permeate/yeast extract medium with immobilized *Lactobacillus helveticus*in a two-stage process: Model and experiments, Enzyme Microbiol Technology, 38: 324-337.

[30] Shaba, B.O. (2000) Effect of He-Ne and Nitrogen on the Viability of E. coli and it genetic content, M.Sc. Thesis, College of Science, Baghdad University. 135 P.

[31] Simeth, S. (1985) Biochemistry with Clinical Correlation. 3rd Ed. N.Y.

[32]Singh, J. and Ranganathan, B. (1977) Acid production and prokaryotic activity in milk by gamma irradiation induced mutants of lactobacilli. Journal of Dairy Science. 39, 1611-1612.

[33] Teuber, M. (1993). Lactic acid bacteria In " Biotechnology" 2nd ed. Edited by H. J. Rehm and G. Reed in cooperation with A. Puhler and P. Stadler Sweden.

[34] Teuber, M. (1999) Fermented milk products, in "The Microbiological Safety and Quality of Food" Chap. 23. Microbiological Safety and Quality of food Vol.1, 535-589.

[35] Teuber, M.andErko, S. (1988) Molecular taxonomy and phylogenetic position of lactic acid bacteria. Biochemistery. 70: 317-324.

[36] Tiwari, K. P., Pandey, A. and Mishra, N. (1979) Lactic acid from molasses by mixed population of lactobacilli. Zb. Bakt II 134: 544-546.

[37] Vodnar, D.C., J. Venus, R. Schneider and C. Socaciu, (2010). Lactic acid production by Lactobacillus paracasei 168 in discontinuous fermentation using lucerne green juice as nutrient substitute. Chemistery and Engineering Technolology, 33: 468-474.

[38] Xuejuan G. and Da X. 2009. Molecular mechanisms of cell proliferation induced by low power laser irradiation. Journal of Biomedical Science. 16:4

CHAPTER THREE

THE PROTECTIVE EFFECT OF CURCUMIN ON NEPHROTOXICITY INDUCED BY CARBON TETRACHLORIDE IN RATS

Sahar Mohamed Kamal Shams El Dine

Department of Pharmacology, Faculty of Medicine, University of Ain-Shams, Cairo, Egypt.

saharkamal2003@hotmail.com

ABSTRACT

The formation of carbon tetrachloride (CCl4) results in an excessive production of reactive oxygen species (ROS) in many tissues, including the kidney. These Free radicals cause a high degree of renal impairment, possibly even renal failure. The aim of the present study is to investigate the possible beneficial effect(s) of curcumin against nephrotoxicity induced by CCl4 in albino rats. Eighteen rats were divided into three groups (number per each group=6 rats). Group I: The control DMSO-treated group. Group II: The control CCl4-treated group, treated with 0.5 ml of 10% CCl4 dissolved in olive oil, per rat, twice a week for 2 weeks. Group III were treated with the same dose and concentration of CCL4 that was co-administered with curcumin dissolved in DMSO at a dose of 50 mg/ kg ip. The duration of treatment in all groups was 5 consecutive days. At the end of the experiment, blood samples were collected and used for determination of kidney functions; urea, creatinine and creatinine/albumin ratio were measured and analyzed.

Malondialdehyde (MDA), reduced glutathione (GSH), nitric oxide (NO), superoxide dismutase (SOD) and glutathione-S-transferase (GST) were evaluated in kidney homogenate. Rats treated with CCl_4 showed significant elevations in kidney function tests, MDA, GSH, NO but with a significant decrease in the activities of SOD and GST. The treatment with curcumin has preserved the measured kidney function tests close to the control values. It significantly restored the activities of SOD and GST and decreased kidney MDA, GSH and NO levels. Curcumin therapy could provide an interesting nephroprotective effect which might be explained partially via diminishing the generation of MDA and NO together with an increase in the activities of antioxidant enzymes.

Keywords: Curcumin, Nephroprotection, Carbon tetrachloride, albino rats.

INTRODUCTION

Curcumin has been implicated to be neuroprotective in a variety of neurodegenerative disorders such as alzheimer's disease (AD) and cerebral ischemia [1&2]. This neuroprotection directs our research toward further studies on its possible protective effects on other vital organs (such as the kidneys) against toxic substances. [eg carbon tetrachloride (CCl_4).]

CCl_4 is a potent lipid-soluble hepatotoxin bound to a lipid and a protein, used to enhance the peroxidative process [3]. Its toxic effect is dependent on the excessive production of the trichloromethyl radical ($CCl_3\bullet$), which interacts with oxygen to form the more toxic trichloromethyl peroxyl radical ($CCl_3O_2\bullet$) [4]. It can cause the formation of reactive oxygen species (ROS) in many vital tissues such as the liver, kidney, heart, lung, testis, brain, and blood [5]. Toxic free radicals lead to marked lipid peroxidation that results in excessive damage to cell membranes and in the development of a number of pathological changes in renal impairment [6&7]. These toxic renal effects occur via the destruction of renal mitochondrial function including the calcium flux across mitochondrial membranes [8].

Many researchers have tried to prove that several herbs and plant-derived pure molecules could protect organs against CCl4 induced oxidative stress by altering the levels of increased lipid peroxidation, by enhancing the decreased activities of antioxidant enzymes [such as superoxide dismutase (SOD), catalase (CAT) and glutathione-S-transferase (GST)] and by enhancing the decreased level of the reduced glutathione (GSH) [9].

In Alzheimer's disease (AD), where neuronal oxidative stress was found to play an important role in its pathogenesis, Curcumin has been reported to be a good inhibitor of c-Jun N-terminal kinase (JNK) mediated gene transcription [10]. JNK is an important member of the mitogen-activated protein kinase (MAPK) family, which can be activated by a variety of stimuli including neurotoxic insults, environmental stress and apoptotic agents [11&12]. JNK plays an important role in mediating MPTP-induced neurotoxicity and leads to a marked loss of nigrostriatal dopaminergic neurons. So it is reasonable to assume that the blockage of the JNK pathway by curcumin may prevent or effectively slow down the progression of PD (13&14).

On the same line of cellular protection by curcumin against tissue toxins, the aim of the present study is to investigate the possible protective role of curcumin on kidney tissues against CCl4 damage that would be mediated via a process of oxidative stress induced by CCl4.

METHODOLOGY

Chemicals

All chemicals required for all biochemical assays were analytical grade and were obtained from Sigma-Aldrich Chemicals Co., St. Louis, USA.

Animals

Adult male albino rats of Wistar strain weighing 200-250 g were provided by the animal House of El Messiry Institute of Medical Research, Faculty of Medicine, Ain Shams university. Cairo, Egypt, was used throughout this study. The rats had free access to food and water *ad libitum* and were maintained in a controlled environment under standard conditions of temperature and humidity with an alternating 12 hr light and dark cycle.

Animal grouping

Eighteen rats were divided into three groups (number per each group=6 rats):

Group I: control DMSO-treated group.

Group II (control CCl4-treated group): treated with 0.5 ml of 10% CCl4 dissolved in olive oil per rat twice a week for 2 weeks, in accordance to Lin et al [15].

Group III: treated with the same dose and concentration of CCL4 with curcumin dissolved in saline at a dose of 50 mg/ kg ip. The duration of treatment in all groups was 5 consecutive days. At the end of the experiment, blood samples were collected and used for determination of kidney functions. Urea, creatinine and creatinine/albumin ratios were measured and analyzed.

Malondialdehyde (MDA), reduced glutathione (GSH), nitric oxide (NO), superoxide dismutase (SOD) and glutathione-S-transferase (GST) were evaluated in kidney homogenate.

It should be noted that in a pilot study, six rats were injected with olive oil alone, another 6 rats were injected with a mixture of DMSO and curcumin, and 6 more rats were injected with a mixture of olive oil, DMSO and curcumin for 5 consecutive days. There were no changes in the measured parameters compared to the group I control DMSO-treated group.

Ethics

The protocol of the study was approved by the Animal Ethics Committee of Faculty of Medicine, Ain Shams University. The study was conducted in accordance with the NIH guide for Animal Welfare.

All groups were anesthetized with urethane (1 g/kg ip) at the end of the 5th day and blood samples were collected from each rat through retro-orbital puncture. The serum was separated and stored at -80* for biochemical analysis. Next, the rats were sacrificed, and kidney tissue was rapidly removed and homogenized in phosphate buffer saline (0.1 M PBS, pH 7.4). The homogenates were centrifuged at 10000 rpm for 30 min at 4 °C, and supernatants were stored at −70 °C *unl. At that time, biochemical assays could be performed.

Biochemical tests

-1- Kidney function test:

Serum urea and serum creatinine levels were assayed in the samples by a colorimetric method [16] using commercial diagnostics kits (Diamond Diagnostics, Egypt). The levels of serum urea and creatinine were expressed in mg/dL. The serum albumin level was assessed calorimetrically using commercial diagnostic kits obtained from Diamond Diagnostics, Egypt. Next, the creatinine /albumin ratio (C/A) was calculated from the results obtained. All serum parameters were measured using the Spectrophotometer 1200 (UNICO Instruments.inc.USA).

-2-Determination of lipid peroxide level:

Lipid peroxidation level in the kidney homogenate was determined as thiobarbituric acid reactive substances (TBARS) by measuring malondialdehyde (MDA) level spectrophotometrically in kidney homogenates [17]. Briefly, 0.5 ml supernatant of kidney homogenate was mixed with 0.6% thiobarbituric acid (TBA) solution in water and 1% aqueous orthophosphoric acid solution, and heated in a boiling water bath for 45 min. The pink-colored chromogen formed by the reaction of TBA with MDA was extracted by n-butanol and measured at 535 nm. The results were expressed as MDA nmol/mg tissue protein.

-3- Nitric oxide (NO) level in kidney homogenate:

Nitric oxide level in kidney tissue homogenates supernatant was determined by its reaction with Greiss reagent. The assay is based on the diazotization of sulfanilic acid with nitric oxide at acidic pH (2-3) and subsequent coupling with N-(10-naphthyl)- ethylenediamine to yield an intensely pink colored product that is measured spectrophotometrically at 543 nm. Nitrite level was calculated using a standard curve for sodium nitrite and its level was expressed in umol/mg tissue protein [18].

-4- Glutathione (GSH) in kidney:

Levels of GSH in kidney homogenates were assayed [19]. The deproteinzation of kidney homogenate was achieved by using 10% trichloroacetic acid and centrifuged at 3500 rpm for 10 min. 200µl supernatant was mixed with dipotassium hydrogen phosphate buffer (PH 8) and 0.4% 5,5'-dithiobis 2-nitrobenzoic acid (DTNB) solution. The yellow-colored substance formed by the reaction of GSH and DTNB was measured at 412 nm. The results were expressed as GSH mg/g tissue.

-5- Antioxidant enzymes in kidney:

a. Superoxide dismutase enzyme (SOD) activity:

Superoxide dismutase activity in kidney homogenate supernatant was determined spectrophotometrically according to the method by Roth & Gilbert, [20]. This method is based on measuring the % inhibition in the auto-oxidation of pyrogallol in the presence of SOD enzyme. One unit of SOD represents the amount of enzymes required to inhibit the rate of pyrogallol oxidation by 50% at 25°C. The activity was expressed as units/mg protein.

b. Glutathione-S-transferase enzyme (GST) activity:

GST activity in kidney homogenate supernatant was determined spectrophotometrically by the method described by Habig et al [21]. The conjugation of 1- chloro- 2, 4- dinitrobenzene (CDNB) with reduced glutathione was measured. The conjugation is accompanied by an increase in absorbance at 340 nm where the rate of increase in the absorbance was directly proportional to the GST activity in the homogenate. The results were expressed in U/min/mg of protein.

-6- Determination of tissue protein content:

Protein content in kidney homogenate was determined according to Lowry's method. [22], using bovine serum albumin (BSA) as a standard.

-7- Statistical analysis:

All of the data obtained was expressed as a mean ± SD. Statistical analysis was performed by a one-way analysis of variance (ANOVA) followed by Tukey's test of significance, as a post-hoc test using GraphPad Prism (Graph software company, version 5.00). $P<0.05$ was considered statistically significant.

RESULTS

Table 1: Effect of 5 day co-administration of CCl_4 and curcumin and the kidney function tests.

The experiment shows a significant ($p<0.05$) increase in all kidney function tests in the CCl_4-treated group compared to the control non-treated group I. The concomitant administration of curcumin with CCl_4 protects against the impairment of these tested parameters.

Table (1): Data is presented as mean ± SD, n=6 rat/group.

* $P<0.05$= significant increase in CCl_4 -treated group II compared to group I (control group) **

$P<0.05$= significant decrease in CCl_4+ curcumin -treated group III compared with group II (CCl_4-treated group).

Table 1

Parameter	Control	CCl4-treated	CCl4+curcumin
Serum Albumin (gm/dL)	3.0±0.1	7.6 ±1.2*	3.4±0.13**
Serum Creatinine (mg/dL)	0.9±0.009	4.5±0.2*	1.0±0.01**
Creatinine/Albumin ration	0.3± 0.09	0.59±0.17	0.29±0.08**

Table 2: Effects of 5 day co-administration of CCl4 and curcumin on MDA (in nmol/mg tissue protein), GSH (in mg/g tissue protein) NO (in µmol/mg tissue protein), SOD (as IU/ mg tissue protein) and GST (as IU/min/mg tissue protein).

This table shows a significant ($p<0.05$) increase in MDA and NO in kidney homogenates of rats treated with CCl4 compared to the control non-treated group I. Concomitant administration of curcumin with CCl4 decreases both oxidative parameters in kidney homogenates, but at the same time, it increases the activities of the measured antioxidant enzymes (SOD, GST) in kidney homogenates.

Table (2): Data is presented as mean ± SD, n=6 rat/group.

* $P<0.05$= significant increase in CCl4-treated group II compared to group I (control group).

ᵗ P<0.05= significant decrease in CCl4-treated group II compared to group I (control group).

** P<0.05= significant decrease in CCl4+ curcumin -treated group III compared with group II (CCl4-treated group).

ᵗᵗ P<0.05= significant increase in CCl4+ curcumin -treated group III compared with group II (CCl4-treated group).

Table 2

Parameter	Control	CCl4-treated	CCl4+curcumin
MDA	0.12±0.01	5.4±1.2*	0.14±0.02**
GSH	3.4±0.02	0.2±0.01ᵗ	4.6±0.14ᵗᵗ
NO	4.5±0.4	22.34±2.3*	2.6±0.6**
SOD	85.5±5.5	4.7±1.6ᵗ	120.5±20.23ᵗᵗ
GST	20.56±3.2	1.23±0.09ᵗ	25.66±2.45ᵗᵗ

DISCUSSION

The results of the present study revealed a nephroprotective effect of co-administration of curcumin with CCl4 against CCl4 induced renal toxicity in albino rats. CCl4, when administrated, is distributed and deposited to organs such as the liver, brain, kidney, lung and heart. It is converted by cytochrome P-450 into the reactive metabolites, trichloromethyl radical (•CCl3) and trichloromethyl peroxide radical (CCl3O2•). These free radicals initiate the peroxidation of membrane poly unsaturated fatty acids, cell necrosis, GSH depletion, membrane damage and loss of antioxidant enzyme activity [23].

It has also been reported that systemically administered CCl_4 in rats was distributed at higher concentrations in the kidney than in the liver [24].

Studies have revealed that curcumin inhibits JNK, which is a tyrosine kinase enzyme inhibitor, so preventing the phosphorylation of Bcl-2/Bax heterodimer, that in turn protects tissues from apoptotic death as shown by the western blot and the electron microscope. By the same mechanism, it was found that curcumin prevents mitochondrial dysfunction, due to either environmental or genetic factors via its protection of the JNK enzyme [25&26].

Curcumin could attenuate the oxidative stress-induced by CCl_4, increase the activities of antioxidant enzymes, and prevent nephrotoxicity by CCl_4 via its possible anti-oxidant beneficial effects. Furthermore, curcumin keeps the mitochondria integrity and preserves mitochondria which normally is considered to possess a critical role in apoptosis and have a damaging effect in a model of Parkinson's disease (PD) induced by a toxic substance, MPTP . Its cytoprotective effect can be induced by its inhibition of the activation of JNKs mitochondria pathway induced by the MPTP lesion and this effect can be inferred on the model in the present study. As in MPTP-induced PD, it prevents dopaminergic neurons from degeneration following MPTP insult by inhibition of neuronal phosphorylation and its oxidative damage that usually leads to neuronal apoptosis [27&28].

CONCLUSION

Thus, the protective effect of curcumin shown in the present study is clinically relevant. There was significant nephroprotection generated when we administered curcumin, co-administered with CCl4. Furthermore, from the results obtained here, we can conclude that curcumin possess an anti-oxidative activity that may be of therapeutic benefit for renal patients.

REFERENCES

[1] Wang Q, Sun AY, Simonyi A, Jensen MD, Shelat PB, Rottinghaus GE, MacDonald RS, Miller DK, Lubahn DE, Weisman GA, Sun GY. Neuroprotective mechanisms of curcumin against cerebral ischemia-induced neuronal apoptosis and behavioral deficits. J Neurosci Res, 82:138–148.(2005).
[2] Yang F, Lim GP, Begum AN, Ubeda OJ, Simmons MR, Ambegaokar SS, Chen PP, Kayed R, Glabe CG, Frautschy SA, Cole GM.: Curcumin inhibits formation of amyloid beta oligomers and fibrils, binds plaques, and reduces amyloid in vivo. J Biol Chem, 280:5892–5901.(2005)
[3] Cabr'e M, Camps J, Paternáin JL, Ferré N, and Joven J. Time- course of changes in hepatic lipid peroxidation and glutathione metabolism in rats with carbon tetrachloride-induced cirrhosis, Clinical and Experimental Pharmacology and Physiology, 27, 694–699.(2000)
[4] Behar-Cohen FF, Heydolph S, Faure V, Droy-Lefaix MT, Courtois Y and Goureau O. Peroxynitrite cytotoxicity on bovine retinal pigmented epithelial cells in culture, Biochem Biophys Res Commun, 226, 842-849.(1996).
[5] Jayakumar T, Sakthivel M, Thomas PA and Geraldine P, Pleurotus ostreatus. an oyster mushroom, decreases the oxidative stress induced by carbon tetrachloride in rat kidneys, heart and brain, Chemico-Biological Interactions, 176, 108–120. (2008)

[6] Khan MR, Rizvi W, Khan GN, Khan RA and Shaheen S. Carbon tetrachloride-induced nephrotoxicity in rats: protective role of Digera muricata, Journal of Ethnopharmacology, 122, 91–99.(2009)
[7] Khan RA, Khan MR, Sahreen S and Bokhari J,. Prevention of CCl4-induced nephrotoxicity with Sonchus asper in rat, Food and Chemical Toxicology, 48, 2469–2476.(2010)
[8] Natarajan SK, Basivireddy J, Ramachandran A, Thomas S, Ramamoorthy P, Jacob M et al,. Renal damage in experimentally-induced cirrhosis in rats: role o oxygen free radicals, Hepatology, 43, 1248–1256.(2006).
[9] Rajesh MG and Latha MS. Protective activity of Glycyrrhiza glabra Linn. on carbon tetrachloride-induced peroxidative damage, Indian J Pharmacol, 36, 284-287.(2004).
[10] Collett GP, Campbell FC. Curcumin induces c-jun N-terminal kinase-dependent apoptosis in HCT116 human colon cancer cells. Carcinogenesis, 25:2183–2189.(2004)
[11] Huang C, Rajfur Z, Borchers C, Schaller MD, Jacobson K. JNK phosphorylates paxillin and regulates cell migration. Nature 424:219–223.(2003).
[12] Becker EB, Bonni A. Pin1 mediates neural-specific activation of the mitochondrial apoptotic machinery. Neuron 49:655–662.(2006).
[13] Saporito MS, Brown EM, Miller MS, Carswell S. CEP-1347/KT-7515, an inhibitor of c-jun N-terminal kinase activation, attenuates the 1-methyl4-phenyl tetrahydropyridine-mediated loss of nigrostriatal dopaminergic neurons In vivo. J Pharmacol Exp Ther. 288:421–427.(1999)
[14] Wang W, Shi L, Xie Y, Ma C, Li W, Su X, Huang S, Chen R, Zhu Z, Mao Z, Han Y, Li M. SP600125, a new JNK inhibitor, protects dopaminergic neurons in the MPTP model of Parkinson's disease. Neurosci Res, 48:195–202.(2004)
[15] Lin HM, Tseng HC, Wang CJ, Lin JJ, Lo CW and Chou FP. Hepatoprotective effects of Solanum nigrum Linn extract against CCl(4)-induced oxidative damage in rats, Chemico- biological Interactions, 171, 283–293.

Fawcett JK and Scott JE, 1960. A rapid and precise method for the determination of urea, J Clin Pathol, 13, 156-159.(2008)
[16] Fawcett JK and Scott G, Borner U, Busch EW and Bablok W. Enzymatic assay of creatinine in serum: comparison with Jaffe methods (author's transl), J Clin Chem Clin Biochem, 17, 683-687.(1979)
[17] Mihara M and Uchiyama M. Determination of malonaldehyde precursor in tissues by thiobarbituric acid test, Analytical Biochemistry, 86, 271–278.(1978).
[18] Green LC, Wagner DA, Glogowski J, Skipper PL, Wishnok JS and Tannenbaum SR. Analysis of nitrate, nitrite, and [15N] nitrate in biological fluids, Analytical Biochemistry, 126, 131–138.(1982).
[19] Ellman GL. Tissue sulfhydryl groups, Arch. Biochem. Biophys, 82, 70-77.(1959).
[20] Roth EF and Gilbert HS. The pyrogallol assay for superoxide dismutase. absence of a glutathione artifact, Analytical Biochemistry, 137, 50–53.(1984).
[21] Habig WH, Pabst MJ and Jakoby WB. Glutathione S-transferases. The first enzymatic step in mercapturic acid formation, The Journal of Biological Chemistry, 249, 7130–7139.(1974).
[22] Lowry OH, Rosebrough NJ, Farr AL and Randall RJ. Protein measurement with theFolin phenol reagent, The Journal of Biological Chemistry, 193, 265–275.(1951).
[23] Ko KM, Ip SP, Poon MK, Wu SS, Che CT, Ng KH and Kong YC. Effect of a lignin-enriched fructus schisandrae extract on hepatic glutathione status in rats: protection against carbon tetrachloride toxicity, Planta Med, 61, 134- 137.(1995).
[24] Sanzgiri UY, Srivatsan V, Muralidhara S, Dallas CE and Bruckner JV. Uptake, distribution, and elimination of carbon tetrachloride in rat tissues following inhalation and ingestion exposures, Toxicology and Applied Pharmacology, 143, 120–129.(1997).
[25] Rush GF, Smith JH, Newton JF and Hook JB. Chemically induced nephrotoxicity: role of metabolic activation, Critical Reviews in Toxicology, 13, 99–160.(1984).

[26] Abraham P, Wilfred G and Cathrine SP. Oxidative damage to the lipids and proteins of the lungs, testis and kidney of rats during carbon tetrachloride intoxication, Clinica Chimica Acta, 289, 177–179.(1999)
[27] Hunot S, Vila M, Teismann P, Davis RJ, Hirsch EC, Przedborski S, Rakic P, Flavell RA. JNK-mediated induction of cyclooxygenase 2 is required for neurodegeneration in a mouse model of Parkinson's disease. Proc Natl Acad Sci U S A 101:665–670.(2004).
[28] Jagatha B, Mythri RB, Vali S, Bharath MM. Curcumin treatment alleviates the effects of glutathione depletion in vitro and in vivo: therapeutic implications for Parkinson's disease explained via in silico studies. Free Radic Biol Med, 44:907–917. (2008).

CHAPTER FOUR

ANALYSIS OF THERMAL PROPERTIES OF MIXED (PP/EPR) / CALCIUM CARBONATES

Noah Pierre Marcel Anicet [1*]; Ayina Ohandha Louis-Max [1]
Eba Medjo Roland[3]; Chabira Salem[2];

[1]*Laboratory of Mechanics, Materials, Structure and Production, University of Douala-Cameroon*

[2]*Mechanics Laboratory (lme), rheology of Thin Films and Environment, University of Laghouat –Algeria*

[3] *Fundamental Physics Laboratory, Department of Physics, Faculty of Science, University of Douala-Cameroon*

Email address: noahpierre@yahoo.fr

ABSTRACT

This work deals with the study of a mixture of the copolymer (PP/EPR) with calcium carbonates socal312, socal322v, Winnofil spm from a thermal point of view. These are marketed by the Solvay trading company. Calcium carbonate is added to the mixture to improve the mechanical properties of the copolymer (PP/EPR). Furthermore, this study highlights on the effect of the addition of this component on the thermal properties of the above-mentioned copolymer. An analysis based on the comparison of firstly the thermal behavior shown by the DSC curves and secondly due to the thermal properties achieved by

---

this technique, has shown the conjugated effect of the processing tool types and the concentration of the respective calcium carbonates (5% and 10%) added to reinforce the copolymer. The results have shown two main effects. Primarily, the addition of either socal312, socal322v or Winnofil spm does not change the melting temperatures and the crystallization temperatures significantly. Secondarily the same observation can be done with different concentrations of these respective additives. In fact an accurate observation of the results presented as tables show that the addition of the calcium carbonates tends to decrease the thermal properties of the copolymer and this is regardless of the type of the mixer used. This represents a double advantage. Calcium carbonates not only reinforce the matrix but also reduce the energy cost of processing. These effects are much more pronounced when using Winnofil spm at a low ratio for reinforcement. These results are consistent with those provided by literature, and have the advantage of providing more DSC curves and detailed tables with the thermodynamic greatnesses achieved by the thermal tests performed on the mixtures of (copolymer PP/EPR mixed with calcium carbonates). It is clear from our work that the addition of the compatibilizer promotes the shift of the crystallization temperature of the mixture to lower temperatures. A second scan tends to slightly correct the temperature shift observed in the first scan. In accordence with many authors, it is observed that the α and β phases coexist in all our blends.

Keywords: DSC, PP/EPR, Socal312, Socal322v, Winnofil spm.

INTRODUCTION

The increase in consumption of PP and its different types of blending makes it a plastic product of wider consumption for the coming years. Thus, polypropylene is extensively used in many areas. However, its use remains limited because of its low mechanical resistance to shocks. The simplest way to overcome this inconvenience (i.e. Low impact resistance) is to mix it with other polymers that are more rubbery, particularly elastomers such as EPR (Ethylene-propylene rubber) [1-4]. However, the addition of EPR particles tends to reduce its tensile properties such as the stress at break and the Young's modulus. Therefore, to reduce these adverse effects nano-rigid fillers such as calcium carbonate ($CaCO_3$) are added to the blend [5]. The addition of EPR decreases the crystallinity of the compounds because of its highly amorphous structure. On a processing point of view, EPR can act as a lubricant, and the capacity of the mixture to withstand deformation is also enhanced. The size of the spherulites decreases with increasing concentration of EPR. In fact, the rate of EPR and the $CaCO_3$ have separate effects on the crystallization and melting of the PP, because they exist as separate and discrete phases in the compound, as reported by Hornsby and Premphet [6].

The melting temperature of the PP phase mixtures is almost independent of the elastomer and calcium carbonate content. It has been shown that the degree of crystallinity of the PP in the mixture PP/elastomer/calcium carbonate is between 50 to 52% [7].

When the $CaCO_3$ concentration increases from 0 to 20%, the crystallization temperature of EPR increases from 102.1 to 108.4 ° C. This indicates that some of the nanocharge particles are embedded in the EPR phase, where they can act as nucleating agents. Several previous studies have focused on the relationship between the phase morphology and the mechanical properties in the systems (PP / EPR / nanofillers) [8-12].

During this last decade, the growing interest by manufacturers in polymer blends is the result of representing the cheapest way to produce new materials with intermediate or higher properties of their components. This progressively becomes a big challenge for the coming years [13]. The main difficulty in compounding polymer blends with competitive properties is their almost general immiscibility. The best theories developed in this research field do not exceed the mixing of two polymers. Therefore, to develop new types of blends for different applications, the blends should contain more than two components. This means they must be combined at different concentrations and processing conditions.

These types of blends are used in many areas such as household appliances, automotive, aeronautics, space, etc. This is why the properties of single plastics are optimized by blending. Nowadays, polymer blends represent about 30% of the world's plastic consumption, and this rate is permanently increasing [14-19].

Nevertheless, the mixtures can show some failures in their mechanical properties. It must be noted that the addition of components such as nano-inorganic fillers can increase the resistance of the system to shock. [20–22].

The addition of mineral fillers which are less expensive than the polymers [23–25] help to lower the cost. Nano-inorganic fillers used in the semi-crystalline polymers are usually talc and calcium carbonate, but the mica and wollastonite can also be used alone [26]. In addition to that nano fillers are used to perform functional roles, such as increasing the rigidity or improving the dimensional stability of the polymers [27]. The mode of reinforcement is specific to the desired property. For example to increase the rigidity, the incorporation of reinforcements such as talc [28], calcium carbonate [29] or carbon black are the most common additive solutions. Another route is the use of manufactured reinforcements such as glass fibers [30] or carbon black. Most of the authors [31–34]consider that nano-calcium carbonate fillers do not alter the crystaline structures; the melting enthalpy is lowered when the particle concentration is increased [35].

The aim of this work is to perform mixtures by means of a micro twin screw internal mixer HAAKE Poly Lab System. The mixtures consist of copolymer (PP/EPR) and socal312, socal322v or Winnofil spm calcium carbonates. Each of them are added at a level of up to 5% and 10% respectively. DSC tests have been carried out on different mixtures and the results have been compared with those of the copolymer (PP/EPR) free of additives. Hence, it becomes possible to identify eventual changes of thermal properties.The macroscopic properties of composite materials depend on the nature of constituents, the volume fraction of their spatial arrangement and the quality of the adhesion between components.

In the following section, the method and the different means used to perform this study are described in details. The results are then discussed with a deep analysis. The last section is devoted to the conclusions.

EXPERIMENTAL WORK

Materials

The various components used to prepare the mixtures are:

The copolymer PP/EPR used as a mixture template which is mainly a combination of two components (propylene and ethylene) in different proportions.

For a long time they have been abusively called "copolymers", while they are in fact mixtures of incompatible bi-phasic polymers.

The PP is a semicrystalline polymer that can contain different types of spherulitic crystals α-type, β-type, γ-type and δ-type. The α-monoclinic spherulitic crystal type is the most stable. Given types of spherulitic crystals can be converted to α-type when appropriate thermodynamic conditions are reached. The most probable crystalline phases that occur during the cooling of the polypropylene are crystals of α-type and β-type. It is shown that spherulites gradually become smaller and more irregular in shape as the content of elastomer increases. It has been observed that the average size of the spherulites decreases from 50 to 40 µm as the elastomeric content increases from 5 to 20% Chou et al. [36].

PP/EPR raw material is supplied as granules. A variety of additives are incorporated into the material to help processing. The copolymer PP/EPR contains 22% of EPR fraction. Blends of polypropylene and EPR elastomers are among the first systems that appeared on the market. These products are mainly used in the injection molding process to produce articles for specific applications. The copolymer PP/EPR displays excellent impact properties, very good adhesion properties with paints and its good flowing properties such as melting makes its processing easier. The combination of all these properties makes it a very attractive product for particular applications where high performances are required, such as car bumpers. Moreover, it shows a high stability toward the UV of solar light (prior to UV stabilizers addition). Its use is thus possible even for an unpainted bumper.

A typical DSC curve of the copolymer PP/EPR used is shown in figure1.

Three types of calcium carbonates or fillers have been used and added to the copolymer (PP/EPR). They are socal312, socal322v, Winnofil spm, marketed by the Solvay company

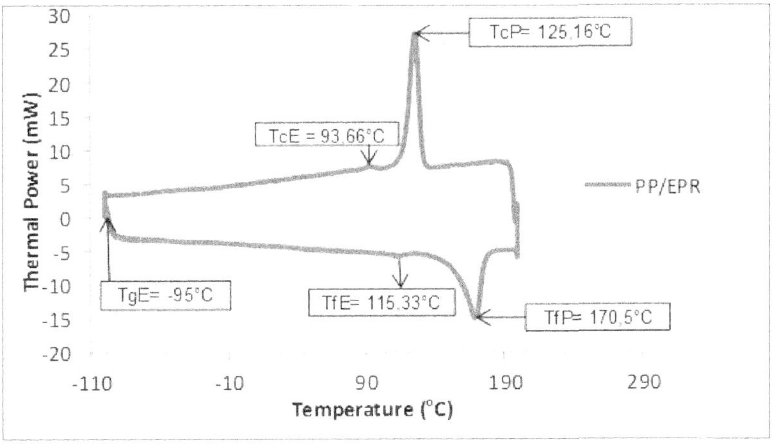

Fig. 1. DSC curve of the copolymer PP/EPR.

The addition of a small amount of calcium carbonates "modified fillers" in the blend at a rate which does not exceed 15% is largely enough to generate physical and chemical interactions required at the interface between the two phases of the mixture, which generates the good impact properties of the material. Socal312 and socal322v are ultrafine, white and odorless. They are used as rheological controller agents in PVC-underbody-car coatings, in printing inks such as pigment-cost saving product leading to rheological control possibilities, and as anti settling fillers in solvent-based paints and lacquers.They provide a high yield value and viscosity, pseudoplastic rheological behaviour, good adherance, possess a very high fineness and roundeal regular crystals and an organic surface treatment. Their properties are very similar. They are shown in table 1 and table 2 below [37-38].

Table 1. socal312 main properties.

Property	**Unit**	**Value**
$CaCO_3$ content	%	98.9
Mean particle diameter	micron	0.07
Specific surface (BET)	m^2/g	19
Loss on drying (at 105°C)	%	0.25
Coating content	%	2.9
Brightness	%	95.8
Yield value	Pa	160
Viscosity	Pa·s	2.4

Table 2. socal322v main properties.

Property	**Unit**	**Value**
$CaCO_3$ content	%	98.8
Mean particle diameter	micron	0.055
Specific surface (BET)	m^2/g	25
Loss on drying (at 105°C)	%	0.4
Coating content	%	3.3
Brightness	%	93.8
Yield value	Pa	180
Viscosity	Pa·s	2.4

The last one is winnofil spm. It is an ultrafine coated precipitated calcium carbonate made by a synthetic crystallization process. It is used as an extender in printing inks to provide improved tint and transparency. It offers a very good dispersibility and enhances the thixotropic properties, improves the slump resistance and rheological control. Winnofil spm reduces the pigment settling in undercoat and heavily filled solvent based paints. Its properties are described in table 3 [39].

Table 3. winnofil spm main properties

Property	Unit	Value
$CaCO_3$ content	%	98.7
Mean particle diameter	micron	<0.1
Specific surface (BET)	m^2/g	20
Loss on drying (at 105°C)	%	0.4
Coating content	%	2.7

The mixtures are prepared in a twin-screw micro and then in an internal mixer HAAKE Poly Lab. The torque variation of the 95/5 and 90/10 mixtures developed in the Micro Twin screw are shown in figure2.

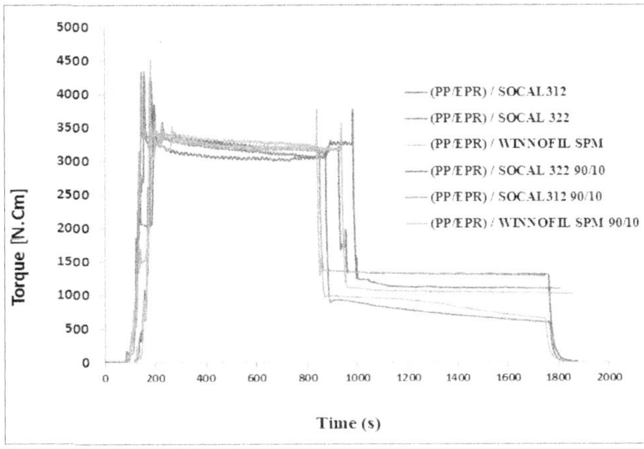

Fig. 2.Torque variation (of 95/5 and 90/10 mixtures) delivered by the Micro Twin screw.

METHODS

Micro twin-screw-type "DSM Micro 15 Compounder" consists of a heating cylinder (maximum temperature 350 °C), in which two screws co-rotate at a contollable speed (5-200 rpm). All settings are made by a Touch screen computer with a multi-function control system (Figure 3). The processing parameters of the micro twin screw to prepare the mixtures are : Mold temperature: 190 °C, Rotating speed: 100 rpm, Weight of raw material : 14 grams for 95/5 and 90/10 blends, the ring is collected after 15 minutes (≈6 grams) and the remainder after 30 minutes, with the aim to study the oxidative degradation of the mixture.

On the internal mixer HAAKE Poly Lab System, the screw rotation and the heating temperature can be controlled The apparatus is piloted by a software "Polylab" used for managing and scheduling information from the experiments (i.e. type of screw, mold temperature, speed screw and holding time). 50 grams of the mixture are processed at a constant speed rotation (40 rpm) and at 180 °C, with a holding time of 20 minutes. Tables 4, 5 and 6 show the compositions of the mixtures obtained to be studied. Figures 2 and 4 show the variations of the torque versus time, delivered by the "rollers" of the micro twin screws and the internal mixer HAAKE Poly Lab System on the mixtures (PP/EPR)/calcium carbonate.

Fig. 3 Midi extruder used to register F = f (t).

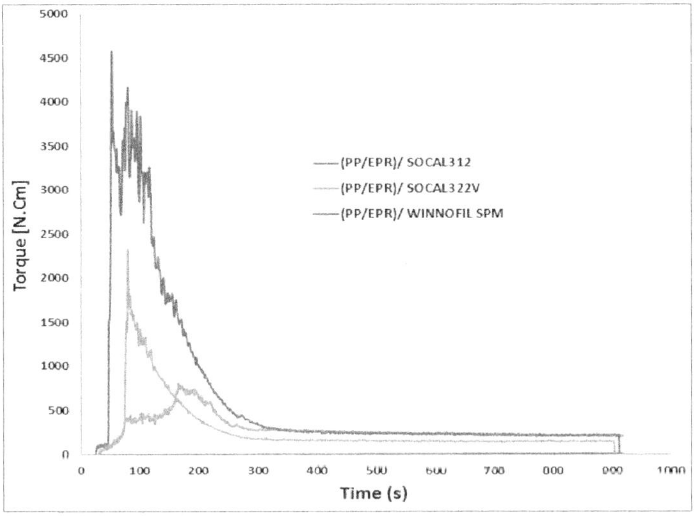

Fig. 4.Torque variation 95/5 mixtures prepared at Haake internal mixer PolyLab System.

Table 4. The 95/5 mixtures prepared on the microtwin screws.

Mixtures	PP/ EPR	SOCAL 312	SOCAL 322V	Winnofil spm
(PP/EPR)/ SOCAL312	95,0%	5,0%	"	"
(PP/EPR)/ SOCAL322V	95,0%	"	5,0%	"
(PP/EPR)/ WINNOFIL SPM	95,0%	"	"	5,0%

Table 5. The 90/10 mixtures prepared on the micro twin screws.

Mixtures	PP/EPR	SOCAL312	SOCAL322V	WINNOFIL SPM
(PP/EPR)/SOCAL312	90,0%	10,0%	"	"
(PP/EPR)/ SOCAL322V	90,0%	"	10,0%	"
(PP/EPR)/ WINNOFIL SPM	90,0%	"	"	10,0%

During the DSC test, the samples are subjected to two thermal cyles (heating/cooling): the first one, from -100 °C to 200 °C with a heating rate of 10 °C/min with an isothermal hold at 200 °C for 2 minutes; the second from 200 °C to -100 °C with a cooling rate of -10 °C/min with an isothermal hold for 2 minutes at -100 °C. The sample temperature is continuously compared to the temperature of a reference. The sample and reference are placed in the same oven whose temperature varies linearly with time, ΔT (temperature difference between the sample and the reference) and then they are measured as a function of the oven temperature. The reference is an empty capsule; the test is carried out in inert atmosphere (argon) to avoid any reaction of the material to be analyzed with oxygen. These measurements are performed thanks to a Mettler-Toledo SA system DSC 30 apparatus. The equipment is calibrated with indium, zinc and lead: the calibration check is done with indium (T_m = 156.6 °C ±0,5 °C; ΔH = 28.45 J/g ± 0,5J/g).

Table 6. The 95/5 mixtures processed in the internal mixer HAAKE Poly Lab System.

Mixtures	PP/ EPR	SOCAL 312	SOCAL 322V	Winnofil spm
(PP/EPR)/ SOCAL312	95,0%	5,0%	"	"
(PP/EPR)/ SOCAL322V	95,0%	"	5,0%	"
(PP/EPR)/ WINNOFIL SPM	95,0%	"	"	5,0%

The evolution of the slope degradation of 95/5 and 90/10 blends, prepared on the micro twin screws, is presented in the folowing table 7.

Table 7. Evolution of the degradation slope of the 95/5 and 90/10 blends prepared on the micro twin screws.

Mixtures	F to 300s	F to 700s	Slope 300-700 s	F to 1000s	F to 1600s	Slope 1000-1600 s
(PP/EPR)/ SOCAL312 (95)/5	3294	3096	-6%	897	647	-28%
(PP/EPR)/ SOCAL322V (95)/5	3116	3032	-3%	1268	1116	-12%
(PP/EPR)/ WINNOFIL SPM (95)/5	3411	3257	-5%	974	720	-26%
(PP/EPR)/ SOCAL312 (90)/10	3354	3183	-5%	1348	1320	-2%
(PP/EPR)/ SOCAL322V (90)/10	3311	3179	-4%	1351	1309	-3%
(PP/EPR)/ WINNOFIL SPM (90)/10	3328	3222	-3%	1109	1046	-6%

Polymer pieces are picked up from the sample (<20 mg of product) and introduced into a 40 µl aluminum capsule with a holed lid. They are then weighed with a Mettler-Toledo balance having an accuracy of ±1 µg. Figure 1 shows the thermal characteristics of the copolymer (PP/EPR).

It is well known that the formation of the crystalline phase in semi-crystalline polymers is related to many factors, such as the thermal history, the molecular characteristics and the impurities. In other studies, it has been shown that fillers such as the nanoparticles of $CaCO_3$ are good nucleating agents that increase the amount of β-type crystals in nanocomposites. The overall crystallinity of these compounds is hardly affected [40-41]. It has also been shown that a rate of about 4 to 6% in volume of $CaCO_3$ nanoparticles favors the formation of the β crystals type, increasing the impact resistance of PP. In the same way, for the same loading rate (i.e. 4-6% in volume) these nanoparticles markedly increase the mechanical resistance of PP. A loading rate of only 4% in volume of $CaCO_3$ nanofillers reduces the activation energy and the melting peak temperature of the α-crystal. These phenomena can be explained by the fact that the crystal growth is preferentially of the β type in PP, in the presence of $CaCO_3$ nanofillers with a concentration of 4 to 6% in volume. As the fraction of $CaCO_3$ nanoparticles increases, the number of crystal nucleons of theβ-type increases as well. However the size of the crystals of β-type decreases, leading to a decrease in the strength properties of the mixtures [40-41]. Most of the authors [42-45] believe that the $CaCO_3$ nanocharges do not change the crystalline structure, however untreated particles interfere with the crystallization process of the polypropylene phase acting as nucleating agents. The $CaCO_3$ particles with the surface-treated do not act as nucleating agents in polypropylene as the crystallinity is not increased.

Calculation of the glass transition temperatures (Tg)

In amorphous statistical copolymers, the Tg is usually a monotonic function of the composition of different copolymers that constitute the mixture[46] and the most commonly used relationship to predict the Tg of copolymer mixtures is the Fox equation[47]. This Fox equation is used by many authors [48,49]:

$$\frac{1}{T_g} = \frac{w_1}{T_{g_1}} + \frac{w_2}{T_{g_2}}$$

where $T_{g1}$and T_{g2} are the glass transition temperatures of the respective polymers (component 1 and component 2) in degrees Kelvin, constituting the comopolymer and (w_1, w_2) are the respective weight fractions.

Another expression of the Fox law is as follow:

$$\frac{L}{T_g} = \frac{W_{g_1}}{T_{g_1}} + \frac{W_{g_2}}{T_{g_2}}$$

So for EPR statistics, Fox law holds, with: T_{gP} = 289K, 173K =T_{gE}, W_P = 42 g/mol, W_E= 28 g/mol, L = W_P + W_E = 70 g/mol. From the T_g of the EPR which is equal to 232K, it is calculated that there are 50% of -E- sequences and 50% of -P- sequences.

RESULTS AND DISCUSSION

The curves of the torque variation of the mixtures (PP/EPR)/Calcium Carbonates socal312, socal322v or Winnofil spm 95/5 and 90/10 exerted on the conical twin screws extruder with a constant speed of (100 rpm) at a temperature of 190 °C and the variation of the mixing torque (PP/EPR)/Calcium Carbonate socal312, socal322v, Winnofil spm 95/5 exerted on the screw rotation at the constant speed (40 rpm) at 180 °C achieved in the internal mixer HAAKE Poly Lab System, are reported in Figures (3, 4) and in Table 3. They show the mixture's degradation slope evolution of the PP/EPR 95/5 and 90/10 copolymer filled with calcium carbonates socal312, socal322v or Winnofil spm processed in the micro twin screw. The DSC curves of the mixtures, (PP/EPR)/Calcium Carbonate socal312, socal322v or Winnofil spm 95/5 prepared in a micro twin screw or in an internal mixer HAAKE Poly Lab System and the 90/10 carried out on micro twin screws are shown in Figures 5,6 and 7, and their thermal properties are listed in Tables 5, 6 and 7. The Changes in the melting enthalpy and crystallization [ΔHmp (J/gPP) and the ΔHcp (J/gPP)] of the studied mixtures produced with the micro twin screws depend on the type and the concentration of the nanofillers incorporarted in the matrix with a rate ranging from 5% to 10% and are shown in Figure 8 (a), (b) and (c).

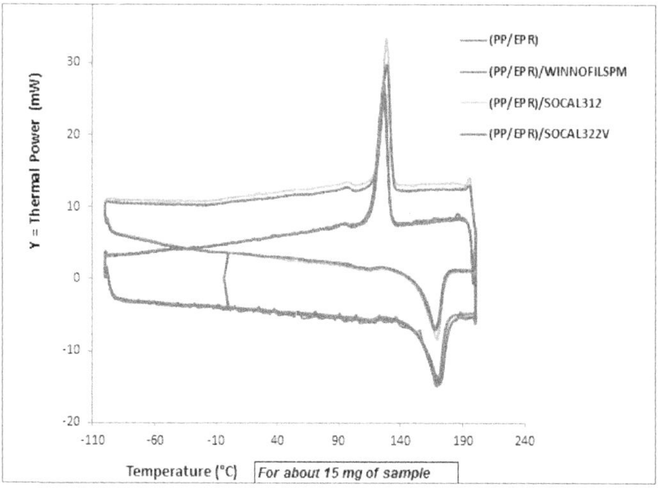

Fig. 5. DSC curves of (PP/EPR)/$CaCO_3$ (95)/5 processed in the micro twin screws.

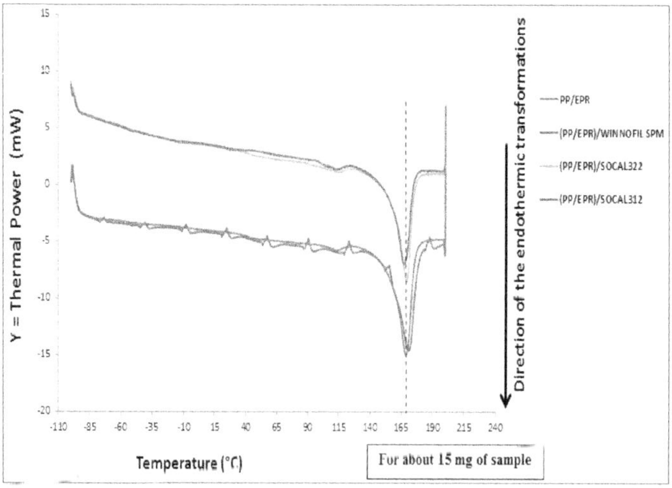

Fig. 6. DSC curves in the direction of the endothermic transformations of Components and (PP / EPR) / Calcium Carbonates (95) / 5 Mixtures developed micro twin screw; for the first thermal cycle.

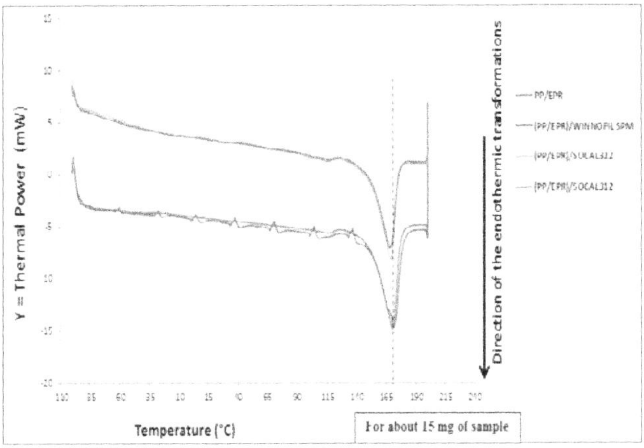

Fig. 7. DSC curves in the direction of the endothermic transformations of Components and (PP / EPR) / Calcium Carbonates (95) / 5 Mixtures developed micro twin screw; for the second thermal cycle.

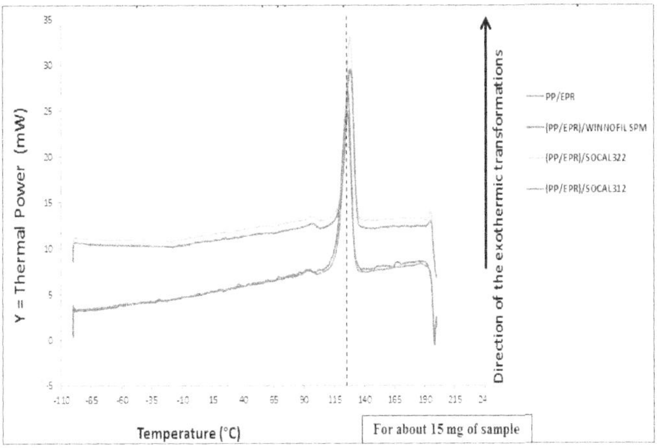

Fig. 8. DSC curves in the direction of the exothermic transformations of Components and (PP / EPR) / Calcium Carbonates (95) / 5 Mixtures developed micro twin screw; for the first thermal cycle.

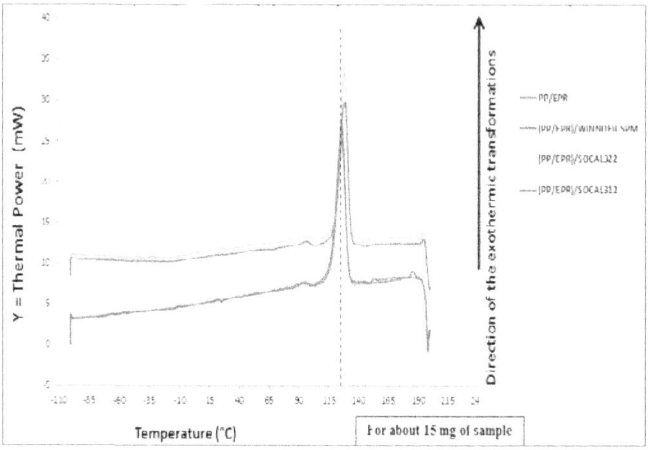

Fig. 9. DSC curves in the direction of the exothermic transformations of Components and (PP / EPR) / Calcium Carbonates (95) / 5 Mixtures developed micro twin screw; for the first thermal cycle.

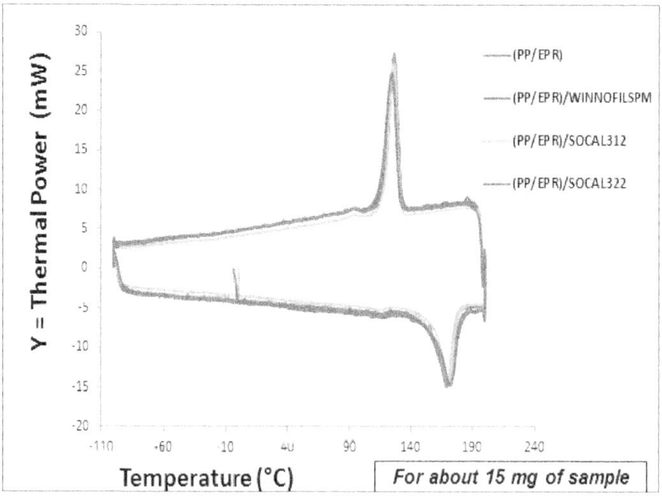

Fig. 10. DSC curves of (PP/EPR)/CaCO3 (90)/10 done on micro twin screw.

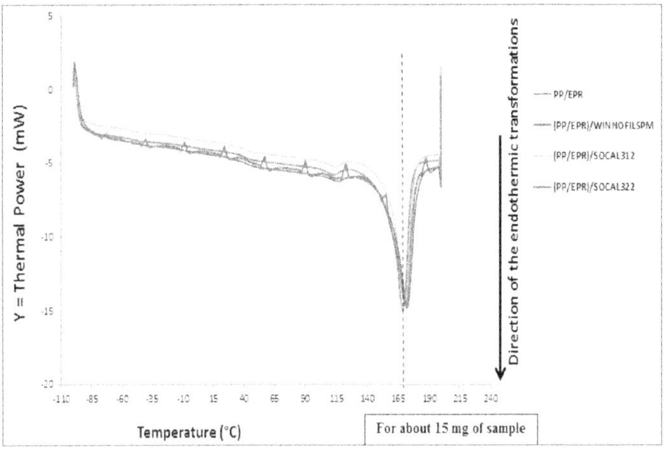

Fig. 11. DSC curves in the direction of the endothermic transformations of Components and (PP / EPR) / Calcium Carbonates (90) / 10 Mixtures developed micro twin screw; for the first thermal cycle.

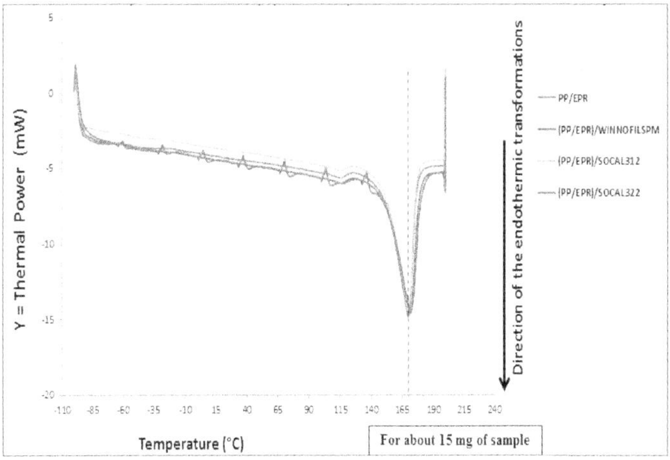

Fig. 12. DSC curves in the direction of the endothermic transformations of Components and (PP / EPR) / Calcium Carbonates (90) / 10 Mixtures developed with the micro twin screw; for the second thermal cycle.

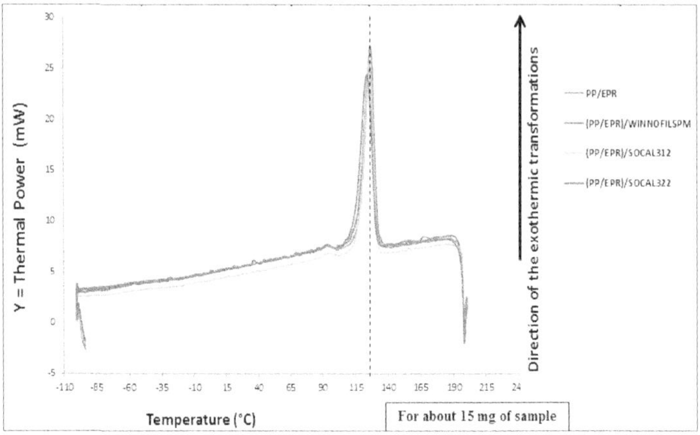

Fig. 13. DSC curves in the direction of the exothermic transformations of Components and (PP / EPR) / Calcium Carbonates (90) / 10 Mixtures developed with the micro twin screw; for the first thermal cycle.

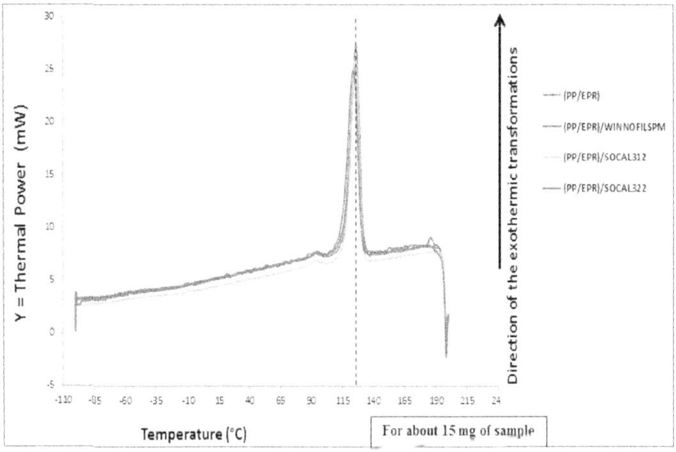

Fig. 14. DSC curves in the direction of the exothermic transformations of Components and (PP / EPR) / Calcium Carbonates (90) / 10 Mixtures developed micro twin screw; for the second thermal cycle.

Table 5. Thermal Characteristics of 95/5 blends prepared on the micro twin screws.

Mixtures	TfE (°C)	TfP (°C)	TCE (°C)	TCP (°C)	ΔHfp (J/gPP)	ΔHCp (J/gPP)
(PP/EPR)	115,33	170,5	93,66	125,16	-103,78	106,17
(PP/EPR)/ SOCAL312	115,66	168,00	96,16	127,66	-100,10	107,90
(PP/EPR)/ SOCAL322V	116,17	169,66	96,17	127,66	98,27	103,52
(PP/EPR)/ WINNOFIL spm	117,00	171,33	95,00	125,33	-92,19	95,83

Table 6. Thermal Characteristics of 90/10 mixtures prepared on the micro twin screws

Mixtures	TfE (°C)	TfP (°C)	TCE (°C)	TCP (°C)	ΔHfp (J/ gPP)	ΔHCp (J/ gPP)
(PP/EPR)	115,33	170,5	93,66	125,16	-103,78	106,17
(PP/EPR)/SOCAL312	115,66	171	92	124,5	-99,79	102,61
(PP/EPR)/SOCAL322V	119,66	172,83	92,66	122,83	96,07	98,60
(PP/EPR)/WINNOFIL spm	-	174	95,83	122	-90,50	100,74

Table 7. Thermal Characteristics of 95/5 blends elaborated on internal mixer HAAKE Poly Lab System.

Mixtures	TfE (°C)	TfP (°C)	TCE (°C)	TCP (°C)	ΔHfp (J/gPP)	ΔHCp (J/gPP)
(PP/EPR)	115,33	170,5	93,66	125,16	-103,78	106,17
(PP/EPR)/SOCAL312	116,66	171,00	95,66	126,83	-95,18	103,18
(PP/EPR) /SOCAL322V	115,16	170,83	95,16	125,83	-93,30	98,36
(PP/EPR)/WINNOFIL spm	-	173,00	93,66	124,33	-95,97	104,85

ANALYSIS

Verification of the Fox law confirms the good characteristics of the copolymer (PP/EPR). This allows us to say that it has correctly been processed by the manufacturer who has fully complied with equal -E- and -P- sequences. Figures 3 and 4 show that the mixtures (PP/EPR)/calcium carbonates socal312, socal322v or Winnofil spm, whatever the fillers rate and the mixer used, were prepared under the same conditions. Indeed the profile of the curves and the torque values are almost similar. For mixtures made in the micro twin-screws, after around 15 minutes of running there is a sudden drop of the torque. This is due to the fact that the ring is collected and shearing becomes important. However this action should have no effect on the thermal degradation rate. However, looking at the results of Table 3, it is found that the thermal degradation of the blends (95/5) made on the micro twin screws is more pronounced at high temperatures with ratios up to 28% in the cases of using socal312. This is inversed with increasing the filler content up to 10%, except for Winnofil spm where the first phenomenon is always observed, but less pronounced.

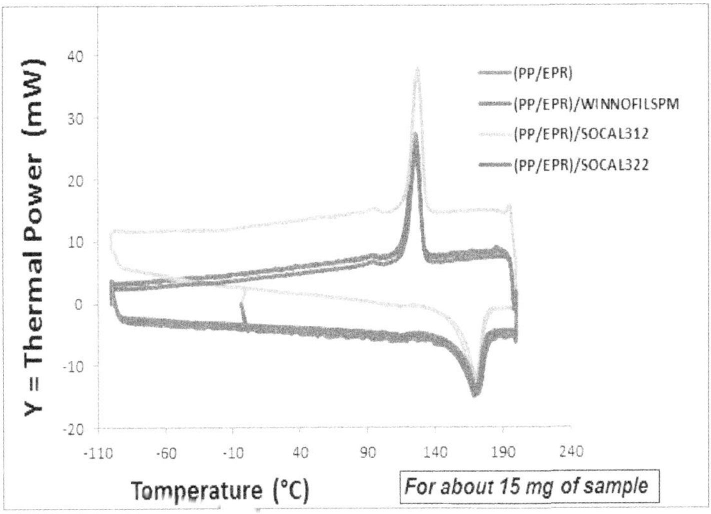

Fig. 15. DSC curves of (PP/EPR)/$CaCO_3$ (95)/5 done on HAAKE Poly Lab System.

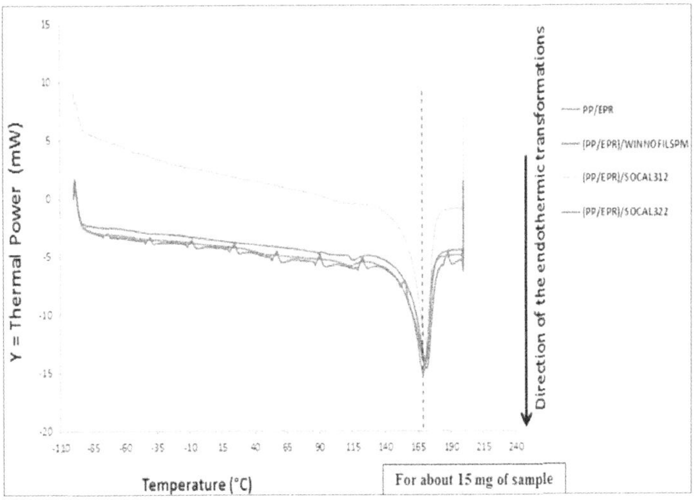

Fig. 16. DSC curves in the direction of the endothermic transformations of Components and (PP / EPR) / Calcium

Carbonates (95) / 5 Mixtures developed on Haake internal mixer Poly Lab System, for the first thermal cycle.

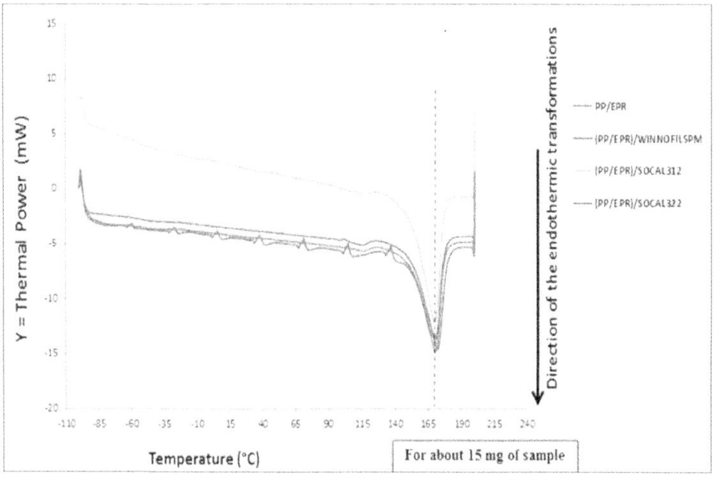

Fig. 17. DSC curves in the direction of the endothermic transformations of Components and (PP / EPR) / Calcium Carbonates (95) / 5 Mixtures developed on Haake internal mixer Poly Lab System, for the second thermal cycle.

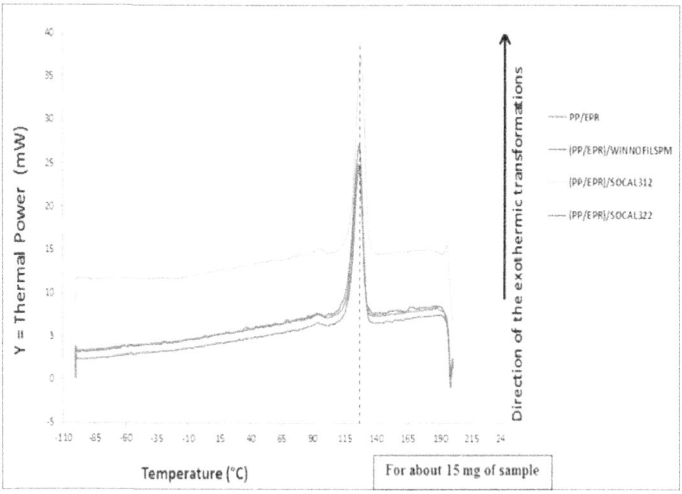

Fig. 18. DSC curves in the direction of the exothermic transformations of Components and (PP / EPR) / Calcium Carbonates (95) / 5 Mixtures developed on Haake internal mixer Poly Lab System, for the first thermal cycle.

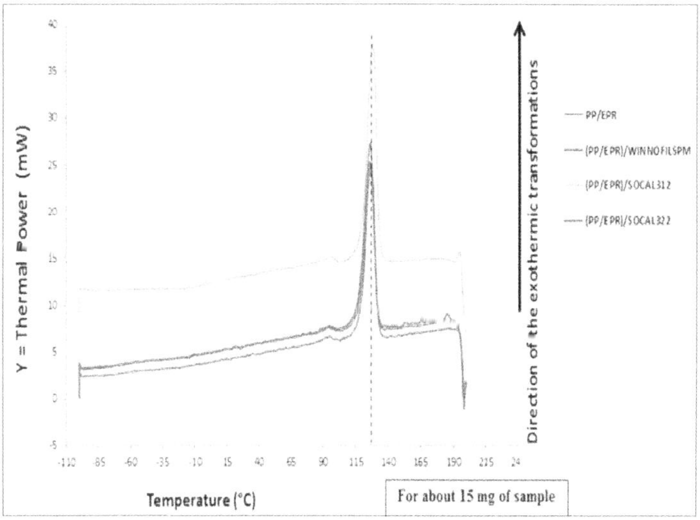

Figure 19. DSC curves in the direction of the exothermic transformations of Components and (PP / EPR) / Calcium Carbonates (95) / 5 Mixtures developed on Haake internal mixer Poly Lab System, for the second thermal cycle.

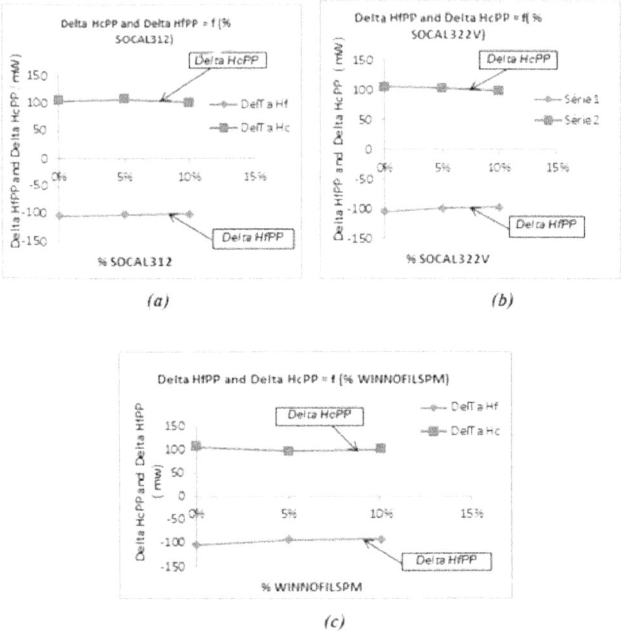

Fig. 20. (a) Evolution of melting enthalpy values, (b) ΔHfp Crystallization ΔHfp (J/gPP), (c) ΔHcp (J/gPP) according to nanofillers rate of 5% to 10% in the copolymer (PP/EPR).

A cursory observation of the DSC curves of the PP/EPR mixtures; (PP/EPR)/socal312; (PP/EPR)/socal322v; (PP/EPR)/Winnofil spm processed on micro twin screws at the rate of (95/5) shows that the curve of Winnofil spm is perfectly superimposed on that of the basic copolymer PP/EPR. Meanwhile the cuives Socal312 and Socal322v are slightly different from that of the copolymer PP/EPR base. However, an increase of only 10% in concentration of the nanofillers cancels completely all the previous differences and they all have the same effect on the basic copolymer. For mixtures prepared on the internal mixer HAAKE Poly Lab System at low levels, only the Socal312 curve is different than that of the basic copolymer PP/EPR. Nonetheless, one observes that whatever the type of mixer used and regardless to the nanofillers rates, the melting temperatures and crystallization of the -P- sequences remain unchanged from the basic copolymer PP/EPR. So we can say that these calcium carbonates can enhance PP/EPR without distorting the melting temperature and crystallization of the -P- sequences, but can change the values of the melting enthalpies and crystallization of the -P- sequences. This observation is supported by the results displayed inTables 5 and 6, where the thermal constants values are almost similar. The Winnofil spm affects the thermal properties of the copolymer (PP/EPR), the drop of the melting enthalpy and the crystallization values of the -P- sequences computed through the integration of their respective peak area reaching 11.6%. This is regardless to the mixing level. It is generally observed for all the mixtures (Figure 8) obtained with the micro twin screws that the addition of the aforesaid calcium carbonates in the copolymer (PP/EPR) decreases the crystallization and melting energies.Tables 5 to 7 present thermal characteristics of the blends. Figures 5 and 10 show the DSC curves of (PP/EPR)/$CaCO_3$ processed on the micro twin screw at (95)/ 5 and (90)/10 respectively. While, figure 15 shows the DSC curves of (PP/EPR)/$CaCO_3$ (95)/5 processed on the HAAKE Poly Lab System. Figures 6 and 7 are the DSC curves in the direction of the endothermic transformations of the components and (PP / EPR) / Calcium Carbonates (95) / 5 for the first and the second cycles. Figures 8 and 9 are the DSC curves in the direction of the exothermic transformations of the components and (PP / EPR) / Calcium Carbonates (95) / 5 for the first and the second cycles.

Figures 11 and 12 display the DSC curves in the direction of the endothermic transformations of the components and the (PP / EPR) / Calcium Carbonates (90) / 10 mixtures develop micro twin screw; for the first and second thermal cycles.

Figures 13 and 14 are reported as the DSC curves in the direction of the exothermic transformations of Components and (PP / EPR) / Calcium Carbonates (90) / 10 mixtures developed by micro twin screw; for the first and the second thermal cycles.

Figures 16 and 17 show the DSC curves in the direction of the endothermic transformations of components and the (PP / EPR) / Calcium Carbonates (95) / 5 Mixtures are developed on Haake internal mixer Poly Lab System, for the first and the second thermal cycles.

Figures 18 and 19 are the DSC curves in the direction of the exothermic transformations of the components and (PP / EPR) / Calcium Carbonates (95) / 5 Mixtures are developed on Haake internal mixer Poly Lab System, for the first and the second thermal cycles.

Figure 20 shows the enthalpy values which change according to the nanofillers rate at 5% and 10% respectively in the copolymer (PP/EPR).

A scrupulous analysis of the results reported in the tables and the DSC curves displayed in the figures allows us to obtain a variety of information and to do some deductions. The most obvious of these are listed below:

First of all, it can be observed that the crystallization and melting temperatures of the -P- sequences are almost similar, whatever the level or the type of calcium carbonate or the type of mixer used. Similarly, the second heat cycle tends to slightly correct the shifts observed during the first heating scan. Indeed, the phases formed depends greatly on the thermal history experienced by the material during the first scan [40-41]. The phenomena of depolymerization and molecular réarengement can explain this effect. It has also been observed that a relatively high concentration of calcium carbonate tends to correct the thermal properties of the copolymer PP / EPR. Therefore, the different DSC curves of the mixtures with higher calcium carbonate concentrations overlap quite correctly. All the above findings are observed for mixtures with a high rate of calcium carbonate performed at Haake PolyLab System, except in the case of using Socal312 where a tacky variation of the enthalpy of crystallization is observed.

It is found that the melting temperature increases for mixtures. This is true for all 95/5 and 90/10 blends performed respectively on Haake PolyLab System and Micro twin screw. It is noted however that Socal312 and Socal322v behave as untreated particles at a low rate especially when using the Micro twin screw mixer.

It is also observed that for different mixtures and rates of additives and whatever the type of mixer used, the surface of the endothermic and exothermic peaks are governed by the values of the melting and crystallization enthalpies of the -P- sequences respectively. Indeed the melting temperatures of all the blends are almost all close to each other, therefore the only parameter that can influence the values of the thermal energy becomes the melting enthalpy of the -P- sequences. The peaks are as great as the enthalpy values are high [40-41].

The study of the thermal properties of the reinforced blends of treated calcium carbonate alter the thermal properties of the mixtures only slightly. The treatment applied to the particle surface at least partially isolated calcium carbonates PP. It made these calcium carbonates unable to play the role of nucleating agents which means that the recovery rate of the particles' fatty acids is significant. The addition of compatibilizer enhances the decrease of mixture crystallization temperature. In the case of these mixtures, the situation is true with a rate of 10% of calcium carbonate regardless to the type of nanofillers. At low rates of nanofillers, the insulating shield stearic acid treatment of calcium carbonate does not have much effects [40-41, 50]. Calcium carbonates play their role of being nucleating agents, excepted for the use of Winnofil SPM at a rate of 5% and mixed on the Haake System PolyLab. Chan et al [40] studied the effect of the addition of the untreated calcium carbonate on the rate of crystallization of PP samples from different percentages of calcium carbonate (particle size of approximately 44nm) by DSC analysis. The results show that the addition of calcium carbonate does not change the rate of crystallinity. Nevertheless, it contributes to the increase of the crystallization temperature by 10 ° C. In the case of our results, the small increase in the crystallization temperature of the -P- sequences with low calcium carbonate content and whatever the mixer used clearly proves the treatment performed on the particles of nanofillers. According to Zeberjad et al[51] who studied the influence of the addition of the untreated calcium carbonate on PP crystallization by DSC analysis, there occurs a significant increase in the crystallization temperature and a decrease of the enthalpy of fusion of the blends due to the nucleating agent (calcium carbonate) in the PP matrix. These results are quite similar to ours where treated nanofillers have been used. The slight difference in the results can be related to the processing conditions of the mixtures and the DSC tests, which could be caused by the superficial destruction of the shield treated nanofillers mixed either on Haake Poly Lab System or on the Micro twin screw. Only the (PP / EPR) / Socal 312 mixture performed on the Micro twin screw is consistent with the results of the authors cited above. However Premphet et al. [52] has shown that in the case of mixtures (PP / EOC) (ethylene-octene copolymer) containing calcium carbonate treated nanofillers, the presence of stearic acid decreases the free enthalpy of the nanofiller's surface compared to the untreated nanofillers and the change in crystallization temperature becomes less significant ($\Delta T = 4^{o}C$) for stearic acid in treated nanofillers ($\Delta T = 22^{o}C$) compared to untreated nanofillers. These results are broadly consistent with what we have found. Differences in the results can be attributed to the complexity of the phenomena related to the composition and properties of the mixtures [51-53].

Some fillers such as $BaSO_4$ [54] and calcium carbonates [55-56] have no effect on the PP crystallinity. The effect of calcium carbonates will be observed only for high rates according to Pukansky Moczo [56]. Some of these results are in perfect correlation with ours where the crystallinity rate is the ratio of sample enthalpy of fusion to a reference enthalpy. It has also been observed that the increase in the calcium carbonate fillers content decreases the melting enthalpies of the -P- sequences mixtures, regardless of the rate of nanofillers and the mixer used. Therefore, a decrease of the crystallinity concentration of the basic PP / EPR coplymer is observed. This result is in good agreement with those results of Denac et al. [57] wherein, the crystallinity degree decreases from 65% to 45% with the addition of talc. Although only slightly or not at all affecting the crystallinity of the α phase of isotactic PP, the calcium carbonate is an element able to promote the formation of β phase when associated with a specific thermal treatment. For example Labour [58] showed that the addition of nanosized calcium carbonates coated with stearic acid may promote the formation of the β phase to a rate of 60%. According to the literature, the crystal phase β would possess different properties compared to those of the α phase [58-66]. Arranz-Andres et al. showed that the γ phase is favored by low cooling rates [67].

Norton and Keller [68] are particularly interested to describe the organization
of crystalline lamellae in different types of spherulites.While, Varga [69] worked on the effects of the thermal cycles on the spherulites form, in particular the growth of spherulites with rings form which are compounds of both α and β phases. In our own case and in accordance with the above observations and table 8 [68, 70], we can deduce the coexistence of the two phases α and β in all our blends

Table 8: The two types of existing spherulites characteristics.

Type	Phase	Crystallisation temperature (Tc) (^oC)
I	α	Tc < 134 ^oC
II	β	110 < Tc < 128 ^oC

Calcium carbonate is industrially used not only to improve the mechanical properties (stiffness, whiteness appearance) but also to reduce the cost of the material. In order to promote its dispersion in the organic matrix, it is generally modified by a surface treatment with the main emplient stearic acid [58, 65], or organosilanes [71-72]. A number of studies analyze its influence on the crystallization of isotactic PP [58, 73-75]. It has been shown that calcium carbonate generates a small nucleating effect [74-75] which causes a small increase in the crystallization temperature (Tc) of the mixtures. This effect could be due to the surface energies of the calcium carbonates [58]. These results are consistent with ours and they are extremly close for lower concentrations of calcium carbonates, and this is regardless to the type of mixer used.The presence of this filler associated with the thermal treatment would promote the development of the β phase. Labour et *al.* [58,65] has shown, in the case of PP with 25% of weight of calcium carbonate treated with stearic acid, that the finished parts contained up to 50% of β-phase. This rate exceeds 10% in the case of the unreinforced polymer. Scientists also found that the greater part is formed only in the presence of stearic acid. These authors put obvious and consistent results reported by others, that the effect of calcium carbonate on the rate of crystallinity or spherulites size is low [74-75].

CONCLUSION

Mixtures of copolymer (PP/EPR) and calcium carbonates socal312, socal322v and Winnofil spm were studied. It appears that calcium carbonates do not alter the melting and crystallization temperatures of the -P- sequences of the mixtures and this is regardless of their filler rates and mixer type used. However, they generally decrease the melting and crystallization enthalpy of the -P- sequences of the matrix which is of a double interest: calcium carbonates reinforce the matrix 2–4 while reducing the energy cost of processing. This phenomenon is much more pronounced when using Winnofil spm. It is observed for the different mixtures and rates of calcium carbonates and whatever the type of mixer used, the intensity and surface of the endothermic and exothermic peaks are governed by the respective melting and crystallization enthalpy values of the -P- sequences. Similarly, the second heat cycle tends to correct the slight shifts observed in the first heat cycle. The addition of compatibilizer enhances the decrease of the mixture crystallization temperature. For these mixtures, this assertion remains true for a rate of about 10% of calcium carbonate and this is regardless to the type of nanofillers used. At low rates of nanofillers, the insulating shield stearic acid treatment of calcium carbonate does not have much effects. In accordence with Varga, Norton and Keller, it appears that α and β phases coexist in all our blends.

REFERENCES

[1] A. Amash and P. Zugenmaier, *Dynamic mechanical investigations on drawn samples of isotactic polypropylene/EPM–Rubber blend*, Journal of Polymer Science Part B: Polymer Physics, Volume 35, Issue 9, 15 July 1997, Pages: 1439–1448.

[2] S. Wong and Y. Mai, *Effect of rubber functionality on microstructure and toughness of impact-modified nylon*

6,6/polypropylene blends: I. Structure – property relationships, Polymer 40 (1999) 1553.

[3] J.Z. Liang, R.K.Y. Li and S.C. Tjong, *Impact fracture behavior of PP/EPDM/glass bead ternary composites*, Polym. Eng. Sci. 40 (2000) 2105.

[4] S. Paul and D. Kale, *Impact modification of polypropylene copolymer with a polyolefinic elastomer*, J. Appl. Polym. Sci. 76 (2000) 1480.

[5] J. Jancar, A.T.J. Dibenedetto and J. Kucera, *Failure méchanics in elastomer toughened PP*, Polymer 34 (1993) 1684.

[6] P.R. Hornsby and K.J. Premphet, *Influence of phase microstructure on the mechanical properties of ternary phase polypropylene composites*, J. Appl. Polym. Sci. 70 (1998) 587.

[7] S.M. Zebarjad , S.A. Sajjadi and M. Tahani , *Modification of fracture toughness of isotactic polypropylene with a combination of EPR and $CaCO_3$ particles*; Journal of Materials Processing Technology, Volume 175, Issues 1–3, 1 June 2006, Pages 446-451

[8] J. Jancar and A.T.J. Dibenedetto, *The mechanical properties of ternary composites of polypropylene with inorganic fillers and elastomer inclusions*, J. Mater. Sci. 29 (1994) 4651.

[9] R. Premphet and P. Horanont, *Phase structure and property relationships in ternary polypropylene/elastomer/filler composites: Effect of elastomer polarity*, J. Appl. Polym. Sci. 76 (2000) 1929.

[10] D. Benderly, M. Siegmann and M. Narksis, *Polymer encapsulation of glass filler in ternary PP/PA-6/glass blends,*Polym. Comp. 17 (1996)86.

[11] S.E. Barbosa, J. Kenny and N.J. Capiti, *Processability and mechanical properties of ternary composites PP/EPDM/GF*, Polym. Comp. 21 (2000) 377.

[12] J. Jancar, *Yielding and impact behaviour of pp/sgf/epr ternary composites with controlled morphology*, J. Mater. Sci. 31 (1996) 3983.

[13] L. Robeson, *Historical Perspective of Advances in the Science and Technology of Polymer Blends*, Polymers 2014, 6, 1251-1265;

[14] G.P. SIMON, *Polymer Characterization Techniques and Their Application to Blends*; Oxford University Press: New York, NY, USA, 2003.

[15] W. Baker, C. Scott, and G.H. Hu, *Reactive Polymer Blending*; Hanser Publishers: Munich, Germany, 2001.

[16] D.R. Paul, and C.B. Bucknall, *Polymer Blends*; John Wiley & Sons: New York, NY, USA, 2000; Volume 1 (Formulation) and Volume 2 (Performance).

[17] L.A Utracki, *Polymer Blends Handbook*; Kluwer Academic Publishers: Dordrecht, The Netherlands, 2002; Volumes 1 and 2.

[18] P. M. A. Noah,L.-M Ayina Ohandja., R. Eba Medjo,S. Chabira and Z. M Ayissi., *Study of Thermal Properties of Mixed (PP/EPR)/Calcium Carbonates*, Indian Journal of Science and Technology, Vol 8(11), 51675, June 2015

[19] L. Utracki,*Commercial polymer blends*. Texte imprime. London: Chapman and Hall; 1998. p. 15–658. ISBN: 978-0-412-81020-6.

[20] MZ Rong, MQ Zhang, YX Zheng, HM Zeng. *Structure-property relationships of irradiation grafted nano-inorganic particle filled polypropylene composites*. Polymer. 2001; 42(1):167–83.

[21] CM Chan, JS Wu, JX Li, YK Chueng,*Polypropylene/calcium carbonate nanocomposites*. Polymer. 2002; 43(10):2981–92.

[22] XC Ren, LY Bai, GH Wang, BL Zhang. *Media plastic conference*. China Plast. 2000; 14:22.

[23] MZ. Rong, MQ. Zang, YX. Zheng, HM. Zeng, K. Friedrich,*Improvement of tensile properties of nano-SiO2/PP composites in relation to percolation mechanism*. Polymer. 2001; 42(7):3301–4.

[24] N. Hasegawa, H. Okamoto, M. Kato, A. Usiki,*Preparation and mechanical properties of polypropylene-clay hybrids based on modified polypropylene and organophilic clay*. J Appl Polym Sci. 2000; 78(11):1918–22.

[25] G. Levita, A. Marchetti, A. Lazzeri,*Polymers:A property database. 2nd ed*. Polym Engng Sci. 1989; 19:39.

[26] B. Pukanszky,*Polypropylene: Structure, blends and composites*. In: Karger-Kocsis J, editor. London: Chapman and Hall; 1995.

[27] Z, Bartczak, AS, Argon, RE, Cohen andM Weinberg,*Toughness, mechanism in semi-crystalline polymer blends: Part 2. High-density polyethylene toughened with calcium carbonate filler particles*. Polymer 1999; 40(9):2347–65.

[28] W. Qiu,K. Mai and H. Zeng,*Effect of silane -grafted polypropylene on the mechanical properties and crystallization behavior of Talc/polypropylene composites*. Journal of AppliedPolymer Science, 2000, vol 77, p 2974-2977.

[29] J. Kucera and J. Kolarik,*Mechanical properties of polyethylene and polypropylene filled with calcium carbonate*. In: Polymer Composites; Sedlacek, B., Ed. Berlin-New York: Walterde Gruyter, 1986, p 545-552.

[30] M.A. Rodriguez-Perez,J.I.Velasco,D. Arencon and J. A. De-Saja,*Dynamic Mechanical Behavior of PP/PET Blends Filled with Glass Beads*. In: Eurofillers 03, Alicante (Spain),2003,

[31] L Bartosiewicz andCG Kelly,*Plastics additives*. Adv Polym Technol. 1987; 7:21–8.

[32] J Menczel andJ Varga,*Influence of nucleating agent on crystallization of polypropylene*. Therm Anal. 1983; 28(1):161–5.

[33] M. Fujiyama and T. Wakino *Role of filler particles on nucleation of polypropylene*. J Appl Polym Sci. 1991; 42:2739–42.

[34] A, Khareh AA, Mitra andS. Radhakrishnan,*Effect of CaCO3 on the crystallization behavior of PP*. J Mater Sci. 1996; 31:5691–702.

[35] WCJ Zuiderduin, C Westzaan, J Hue'Tink andRJ Gaymans. *Toughening of polypropylene with calcium carbonate particles* Polymer. 2003; 44(1):261–75.

[36] J.C. Chou, K. Vijayan, D. Kirby, A. Hiltner, E. Baer, *Ductile-to-brittle transition of rubber-modified polypropylene*, J. Mater. Sci. 23 (1988) 2521.

[37] http://coatings.specialchem.com/product/a-solvay-socal-312#.

[38] http://coatings.specialchem.com/product/a-solvay-socal-322-v

[39] http://coatings.specialchem.com/product/a-solvay-winnofil-spm

[40] C.M.Chan,J.S. Wu,J.X. Li and Y.K. Chueng,*Polypropylene/calcium carbonate nanocomposites*. Polymer, 2002, vol 43, p 2981-2992.

[41] X.C. Ren,L.Y. Bai,G.H. Wang and B.L. Zhang,*Media plastic conference*, China Plast, 2002, vol 14, p 22.

[42] L Bartosiewicz. and C.G. Kelly *Microstructural analyses of homo and copolymer extrusions verified with complimentary techniques*. Advances in polymer technoly, 1987, vol7, p 21-28.

[43] J. Menczel and J. Varga *Influence of nucleating agent on crystallization of polypropylene. Journal of thermal analysis*, 1983, vol 28, p 161-165.

[44] M. Fujiyama and T. Wakino,*Role of filler particles on nucleation of polypropylene*, J.Appl. Polym. Sci, 1991, vol 42, p 2739–2742.

[45] A. Khareh,A. Mitra and S. Radhakrishnan *Effect of $CaCO_3$ on the crystallization behavior of PP*. J. Mater. Sci, 1996, vol 31, p 5691-5702.

[46] A. Eisenberg *Physical properties of polymers*. Washington, DC: ACS; 1984. p. 192–9.

[47] G. Fox *Prediction of glass transition temperatures*. Bull Am Phys Soc. 1956; 1:123

[48] P, Marchese,A Celli,M Fiorini andM. Gabaldi,*Effects of annealing on crystallinity and phase behaviour of PET/PC block copolymers*. Eur Polym J. 2003; 39(6):1081–9.

[49] N. Cameron, JM. Cowie, R. Ferguson, JL. Gomez Ribelles, J. Mas Estelles *Transition from miscibility to immiscibility in blend of poly(methyl methacrylate) and styrene-acrylonitrile copolymers with varying copolymer composition*: A DSC study. Eur Polym J. 2002; 38(3):597–605.

[50] C.G Marisa Rocha, H.M.F.T Antonio. Silva, M.B. Fernanda Coutinho and N. Ana Lucia Silva.,*Study of composites*

based on polypropylene and calcium carbonate by experimental design. Polymer Testing, 2005, vol 24, p 1049–1053.

[51] S.M. Zebarjad,M. Tahani and S.A. Sajjadi,*Influence of filler particles on deformation and fracture mechanism of isotactic polypropylene*. Journal of Materials Processing Technology, 2004, vol 155, p 1459–1464.

[52] K. Premphet and P., Horanont *Influence of stearic acid treatment of filler particles on the structure and properties of ternary-phase propylene composites*. J Appl Polym Sci, 1999, Vol74, p 3445-3454.

[53] J. Kolarik,F. Lednicky,J. Jancar and B. Pukanszky,*Phase structure of ternary composites consisting of polypropylene/elastomer/filler. Effect of functionalized components*. Polymer commun, 1990, vol 31, p 201-204.

[54] K. Wang, J. Wu, L. Ye, and H. Zeng, *"Mechanical properties and toughening mechanisms of polypropylene/barium sulfate composites"*, Composites Part A vol. 34, p. 1199-1205 (2003)

[55] J. Hartikainen, P. Hine, J.S. Szabo, M. Lindner, and T. Harmia, *"Polypropylene hybrid composites reinforced with long glass fibres and particulate filler"*, Composites Science and Technology vol. 65, p. 257-267 (2005)

[56] B. Pukanszky and J. Moczo, *"Morphology and Properties of Particulate Filled Polymers"*, Macromolecular Symposia vol. 214, p. 115-134 (2004)

[57] M. Denac, V. Musil, I. Smit, and F. Ranogajec, *"Effects of talc and gamma irradiation on mechanical properties and morphology of isotactic polypropylene/talc composites"*, Polymer Degradation and Stability vol. 82, p. 263-270 (2003)

[58] T. Labour, *Microstructure et comportement mécanique du polypropylène chargé*, Thèse de Doctorat, Institut National des Sciences Appliquées de Lyon (1999)

[59] M. Aboulfaraj, C. G'Sell, B. Ulrich, and A. Dahoun, *"In situ observation of the plastic deformation of polypropylene spherulites under uniaxial tension and simple shear in the scanning electron microscope"*, Polymer vol. 36, 4 p. 731-742 (1995)

[60] M. Fujiyama, *"Structures and properties of injection moldings of β-crystal nucleator-added polypropylenes. Part 1 Effect of β-crystal nucleator content"*, International Polymer Processing vol. 10, 2 p. 172-178 (1995)

[61] J. Karger-Kocsis and J. Varga, *"Effects of β−α Transformation on the Static and Dynamic Tensile Behavior of Isotactic Polypropylene"*, Journal of Applied Polymer Science vol. 62, 2 p. 291-300 (1996)

[62] S.C. Tjong, J.S. Shen, and R.K.Y. Li, *"Mechanical Behavior of Injection Molded β-Crystalline Phase Polypropylene"*, Polymer Engineering and Science vol. 36, 1 p. 100-105 (1996)

[63] T. Sterzynski, P. Calo, M. Lambla, and M. Thomas, *"Trans- and dimethyl quinacridone nucleation of isotactic polypropylene"*, Polymer Engineering and Science vol. 37, 12 p. 1917-1927 (1997)

[64] T. Sterzynski, M. Lambla, F. Georgi, and M. Thomas. *Studies of the trans-quinacridone nucleation of poly(ethylene-b-propylene)*. Intern. Polymer Processing XII. 1997. Munich: Hanser publishers.

[65] T. Labour, C. Gauthier, R. Séguéla, G. Vigier, Y. Bomal, and G. Orange, *"Influence of the β crystalline phase on the mechanical properties of unfilled and $CaCO_3$-filled polypropylene. I. Structural and mechanical characterisation"*, Polymer vol. 42, 16 p. 7127-7135 (2001)

[66] C. Grein, C.J.G. Plummer, H.H. Kausch, Y. Germain, and P. Béguilin, *"Influence of β nucleation on the mechanical properties of isotactic polypropylene and rubber modified isotactic polypropylene"*, Polymer vol. 43, p. 3279-3293 (2002)

[67] J. Arranz-Andrés, B. Pena, R. Benavente, E. Pérez, and M.L. Cerrada, *"Influence of isotacticity and molecular weight on the properties of metallocenic isotactic polypropylene"*, European Polymer Journal vol. 43, p. 2357-2370 (2007)

[68] D.R. Norton and A. Keller, *"The spherulitic and lamellar morphology of melt-crystallized isotactic polypropylene"*, Polymer vol. 26, p. 704-716 (1985)

[69] J. Varga, *"Supermolecular structure of isotactic polypropylene"*, Journal of Materials Science vol. 27, p. 2557-2579 (1992)

[70] 49. H.D. Keith, F.J. Padden, N.M. Walter, and W. H.W., *"Evidence for a second crystal form of polypropylene"*, Journal of Applied Physics vol. 30, p. 1485-1488 (1959)

[71] Z. Demjen and B. Pukanszky, *"Effect of surface coverage of silane treated $CaCO_3$ on the tensile properties of polypropylene composites"*, Polymer Composites vol. 18, 6 p. 741-747 (1997)

[72] Z. Demjen, B. Pukanszky, and J. Nagy, *"Evaluation of interfacial interaction in polypropylene/surface treated $CaCO_3$ composites"*, Composites Part A vol. 29A, p. 323-329 (1998)

[73] A. Kiss, E. Fekete, and B. Pukanszky, "Aggregation of CaCO3 particles in PP composites: Effect of surface coating", Composites Science and Technology vol. 67, p. 1574-1583 (2007)

[74] B. Pukanszky, K. Belina, A. Rockenbauer, and F.H.J. Maurer, *"Effect of nucleation, filler anisotropy and orientation on the properties of PP composites"*, Composites vol. 25, 3 p. 205-214 (1994)

[75] J. Hartikainen, P. Hine, J.S. Szabo, M. Lindner, and T. Harmia, *"Polypropylene hybrid composites reinforced with long glass fibres and particulate filler"*, Composites Science and Technology vol. 65, p. 257-267 (2005)

CHAPTER FIVE

POSSIBILITY OF RNA EDITING IN CARDIOMYOPATHIES AND CHANNELOPATHIES

Amol P. Jadhav[1], M. Sairam[1], K. J. Princy[2], V. Viswamithra[1], C. Sandeep Kumar[1], Altaf Ali [1], Pratibha Nallari[1]

[1]*Dept. of Genetics, University College of Science, Osmania University, Hyderabad-500007, Telangana, India.*

prathinallari@gmail.com

[2]*College of Technology, Osmania university, Hyderabad500007, Telangana ,India.*

ABSTRACT

RNA editing is an alteration of the RNA by modifications of nucleotides mostly found in the untranslated regions (like Alu repeats) which comprises 10% of the human genome. RNA editing occurs widely in the human transcriptome generating RNA and protein diversity. Cardiac disorders like channelopathies (LQTS, SQTS etc) and cardiomyopathies are multifactorial disorders caused significantly due to mutations. Hypertrophic cardiomyopathy is caused due to 13 mutations mainly in *MYBPC3* and other cytoskeletal genes, while Long QT syndrome is caused due to mutations in sodium and potassium channel genes. Although the overall prevalence of RNA editing and its specific functional impact on many of the cardiac genes is still elusive, these genes contain Alu repeats and hence they may undergo RNA editing leading to functional disruption. Pathogenic mutations in gene/s encoding cytoskeletal proteins have been shown to undergo RNA editing. On the other hand, in cardiomyocytes, micro-RNAs regulate the expression of a number of target genes by pairing to their 3'-UTRs. MicroRNAs such as miR-1, miR-133, miR-194, which play an important role in controlling cardiac gene expression, undergo RNA editing.

Keywords: RNA editing, Cardiomyopathies, Alu repeats, LQT, micro-RNAs

INTRODUCTION

RNA editing is an important epigenetic mechanism involving co/post-transcriptional alteration of RNA sequence encoded by the genome, via insertion, deletion or modifications of nucleotides. This process does not include mechanisms like polyadenylation, splicing or degradation of RNA molecules [1]. Discovery of mitochondrial mRNA modification in kinetoplastid of protozoans led to the coining of the term RNA editing. However, in mammals, it was observed for the first time in mRNAs encoding proteins of the central nervous system.
Editing varies according to the enzyme substrates, machinery, species distributions and molecular mechanisms. It frequently generates protein diversity through changes in start and stop codons, frame shifts, and single-amino acid changes at functionally important positions [2, 3].

All types of RNAs undergo editing and can be grouped into two basic classes: 1) insertion/deletion and 2) substitution [cytosine (C) to uridine (U) and Adenosine (A) to Inosine (I)]. The most prominent RNA editing event in human cells is the substitution of A-to-I, catalysed by members of the double-stranded RNA-specific adenosine deaminase, acting on RNA (ADAR) family of enzymes. Since I in mRNA is recognized as guanosines (G) by

the ribosomes during the course of translation, RNA editing can lead to an altered protein and also play an important role in regulating gene expression [4]. Recently, the apolipoprotein B mRNA-editing enzyme catalytic subunit (APOBEC) family members have been shown to be pivotal in the editing process. The APOBEC family contains a group of cytidine deaminases that include APOBEC-1, -2, -3A, -3B, -3C, -3D, -3F, and -3G, and an activation-induced cytidine deaminase (**AID**) [5, 6]. The best characterized example of editing performed by an APOBEC family member is the editing of the miRNA encoding apolipoprotein B by APOBEC1.
The present review deals with the role of RNA editing in cardiac function and heart related disorders.

Adar Family
The presence of ADARs are likely to be seen in all animals and have been also studied in squids, worms, flies and mammals [7]. Adenosine deaminases are ubiquitously expressed in mammals and are specifically recognized as partially double-stranded (ds) RNA structures such as bulges and loops. ADAR proteins share a common domain structure [8] with a highly conserved catalytic deaminase domain and a variable N-terminal region followed by one or several dsRNA-binding domains (dsRBDs) [9]. The hydrolytic deamination activity of ADARs at the C-6 position of adenosine converts into inosine within the RNA chain [10]. Three members of ADAR family are known in mammalian genomes ADAR1, ADAR2 and ADAR3/RED2 [11]. ADAR1 and ADAR2 are expressed in many tissues, but are most abundant in the brain. ADAR3 is specifically expressed in neuronal tissues [12].

1) Full-length 150-kDa form (ADAR1 p150) with extended N terminus including two Z-DNA/RNA-binding motifs mainly found in cytoplasm and a shorter 110-kDa protein (ADAR1 p110) localized to the nucleus and nucleolus [13].

2) ADAR2: A nuclear localizing protein present in several cells and tissues with arginine enriched N-terminus domain containing a NLS (Nuclear Localization Signal) and an adenosine deaminase domain near the C-terminus. It is the main enzyme responsible for A to I conversions in mammals [14].

3) ADAR3: This enzyme shares ~50% similarity with ADAR2 and is seen only in the neuronal tissues [15-18]. It has an arginine and lysine rich R domain that mediates ssRNA binding and functions as a nuclear localization signal [13].

APOBEC Family

APOBEC (Apolipoprotein B mRNA-editing catalytic polypeptide) are a group of cytidine deaminases, proteins containing one or two zinc binding motifs which catalyse cytidine deamination converting cytosines to uracil on a variety of vertebrate-specific RNA/DNA targets.

APOBEC-2

It is the member of the C-U editing enzyme subfamily which contain 224 amino acid residues, present both in humans and mice. Human APOBEC-2 is encoded on chromosome 6 and is expressed at highest levels in cardiac and skeletal muscle [19, 20]. APOBEC-2 is evolutionarily related to APOBEC-1 and a better conserved gene than APOBEC-1. Like other editing enzymes of the cytidine deaminase superfamily, APOBEC-2 has low, but definite, intrinsic cytidine deaminase activity. The highly tissue-specific distribution suggests that APOBEC-2 may affect editing of RNA in cardiac and skeletal muscle.

All these facts contributed to the deduction that APOBEC-2 may be a novel RNA-editing enzyme with natural substrates in cardiac and skeletal muscle [21]. The identification of APOBEC-2 is important from a functionally evolutionary perspective. Consequently, the substrate(s) for APOBEC-2 and the

significance of its expression in cardiac and skeletal muscle remains elusive.
Computational approaches have revealed that most human genes are subjected to A and to I RNA editing mainly within untranslated regions and introns that contain repetitive elements. These repetitive elements are subdivided into tandemly arrayed (e.g. microsatellites, minisatellites and telomeres) and interspersed elements (e.g. mobile elements viz LINES and SINES and processed pseudogenes).
Alu elements, are short interspersed nuclear elements (SINE's) ~300 bp in length. These elements occur at higher frequency within inter-genic regions, 5' and 3' UTRs, etc., comprising approximately 11% of the human genome. A transcript having two neighbouring Alu repeats in an inverted orientation might form a long hairpin structure acting as an ideal substrate for ADAR enzymes. Recent studies have shown that *Alu* repeats account for over 99% of editing events found so far in humans [22-24].

Cardiomyopathies

Cytoskeletal proteins

Actin is a highly conserved protein which participates in cellular functions like muscle contraction, cytokinesis, cell signalling, cell motility etc. Cardiac actins are encoded by the gene ACTC1. Defects in this gene have been associated with idiopathic dilated cardiomyopathy (IDC) and familial hypertrophic cardiomyopathy (FHC).
Filamins are cytoplasmic protein that crosslink actin filaments. In humans, the filamin family is widely expressed during development but mainly restricted to skeletal and cardiac muscle during adult life. Mutations in the X-linked *FLNA* gene lead to various cardiac defects causing abnormalities in the left ventricular outflow tract.
One RNA editing site has been identified within FLNA transcript that results in the substitution of the amino acid Q to R. This editing region is predicted to form a 32 bp long dsRNA structure

which is vulnerable for most of the RNA editing events. When actin interacts with filamin there is a possibility that the RNA editing event in this site may interfere with the binding of actin to filamin. This may impair its function leading to abnormalities such as FHC and IDC respectively.

HCM is the most common genetic heart muscle disease and the most common cause of sudden cardiac death in the youth with a prevalence of 0.2%. HCM is frequently caused by mutations in genes encoding sarcomeric proteins, especially *MYBPC3*. To date, more than 13 gene mutations, coding for sarcomeric proteins, have been found in patients with HCM. However, a new mutational mechanism for HCM was identified [25]. It was identified that this mutation leads to multi-exons C-terminus deletion in the cardiac myosin binding protein C (*MYBPC3*) gene. The 3.5 kb terminal deletion is mediated by the Alu-repeat element which is a preferable site for ADARs, hence it can be predicted that RNA editing may be playing a role in C-terminus deletion of *MYBPC3*, a key player in HCM associated heart failure.

Dilated Cardiomyopathy; a common cause of congestive heart failure characterized by ventricular systolic dysfunction and ventricular dilation. Idiopathic (25-30%) and familial (30-48%) dilated cardiomyopathies are known with reported mutations in the sarcomere and cytoskeletal genes (**26**). More than 30 genes have been identified that contribute to DCM [27], of these *MYH7*, *ACTC1*, *DES*, *DMD* genes have Alu repeats in their gene structure. As it is known that Alu repeats account for most of the RNA editing events, a possibility of RNA editing occurrence in these genes can be substantiated.

Channelopathies

Channelopathies occur when one of the proteins forming the channels does not function properly altering the electrical properties of the patient's heart. Defects in Na^{+}, K^{+} and Ca^{+2} channels due to genetic mutation or other malfunction leads to cardiac channelopathies such as Long QT Syndrome. Ion channels and neurotransmitter receptors in mammals are well-known targets for A – I RNA editing by ADARs. Some important examples include a number of glutamate-gated ion channels, the GABAA receptor, the voltage-gated Kv1.1 potassium channel, etc. In most of the cases, A – I editing occurs at specific and functionally important regions in the protein, changing key amino acid residues crucial for the protein's function.

Calcium channel

The **L-type calcium channel**, a type of voltage-dependent calcium channel, mediates the release of Ca^{2+} from the sarcoplasmic reticulum into the cytoplasm and thereby plays a key role in triggering cardiac muscle contraction. α1 of VDCC subunit is encoded by different genes like *CACNA1C*, *CACNA1D* , *CACNA1F* and *CACNA1S*. These genes encode different subunits from the main α1 subunit which are $Ca_v1.2$, $Ca_v1.3$, $Ca_v1.4$ and $Ca_v1.1$ respectively. In cardiac myocytes, calcium release is triggered by increased Ca (2+) levels due to activation of the L-type calcium channel *CACNA1C*.

The Ca_V 1.2 channel is broadly expressed in the surface membranes of cardiac, smooth muscle, nerve and neuroendocrine cells by the generation of Ca^{2+} action potentials. The broad expression and importance of the Ca_V 1.2 channel was recently demonstrated by a human mutation that causes Timothy syndrome, a disorder characterized by arrhythmia, autism, and other problems. It has been shown that *CACNA1D* has two novel RNA editing sites. These two editing sites occur in the calmodulin-binding part of the IQ domain and are just 8 bp apart. They convert the amino acids Ile1624 and Tyr1627 into Met and Cys, respectively. These two residues and their vicinity are conserved across metazoan species and are known to play an important role in calcium- dependent inactivation and facilitation of the channel. These two RNA editing positions function to expand the repertoire of protein products generated by this gene and they consequently alter cardiovascular disorders.

Sodium channel

Sodium channels play an essential role in generating the action potential in eukaryotic cells and their transcripts, especially those in insects. They undergo extensive A-to-I RNA editing. At least nine different mammalian sodium channel α-subunits ($Na_v1.1$ to $Na_v1.9$) have been identified. These sodium channel isoforms exhibit distinct expression patterns in the nervous system, skeletal and cardiac muscle. Studies have revealed three RNA editing sites, in the *para* transcript of Drosophila, that encode for major voltage-activated Na^+ channel polypeptide. Differences in electrophysiological properties have been observed among different mammalian sodium channel α-subunit isoforms, which most likely contribute to their specialized functional roles in various cell types. Two U-to-C editing sites in $BgNa_v1$-1 and two A-to-I editing events in $BgNa_v1$-2 have been identified in German cockroach sodium channel gene. Although the impact of RNA editing on sodium channel function remains elusive, it can be predicted that RNA editing may play an important role in altering the protein's function or structure, via the conformational change of the sodium channel, resulting in the disruption of its function.

Potassium channel

Voltage-gated potassium channel plays an important role in the repolarization phase of the action potential. It has been reported that the mRNA transcript of human $K_v1.1$ undergoes A-to-I RNA editing, resulting in an amino acid change from isoleucine to valine (GTT) [28, 29]. Interaction of an inactivating particle with the pore of hKv1.1 channel results in channel inactivation. The I-V amino acid change may disrupt this interaction, resulting in a faster recovery from inactivation. In squid optic lobe, RNA editing generates diverse transcripts encoding functionally diverse kv2 potassium channels. The editing in these regions altered the rates of channel closure with slow inactivation. Similarly, in LQT syndrome, RNA editing of these channel encoding genes may slow down or disrupt their function. Hence it is hypothesized that dysfunction in potassium channels, due to RNA editing, is feasible in LQT syndrome.

tRNA editing

tRNAs can also be potential substrates for the adenosine deaminase enzymes. Modification of adenosine to inosine is shown in some tRNAs. It is known that in all eukaryotes, inosine is present at position 37 (3P adjacent to the anticodon) of tRNA-Ala1 and at position 34 (the first base of the anticodon) of seven or eight different tRNA species in yeast and higher eukaryotes, respectively. Inosine at position 34 (I34) is believed to play a crucial role in protein synthesis by allowing alternative base pairing with U, C or A in the third position of codons. The deaminase enzyme is called scADATs in yeast while ADATs in humans.

Infantile cardiomyopathies are devastating fatal disorders with mitochondrial dysfunction as a common cause. Mutations in the mitochondrial tRNA for isoleucine and nuclear mutations in *SCO2*, *TAZ* and *TMEM70* have previously been identified in disorders with early-onset mitochondrial CMP. Novel mutation has been identified which is a homozygous missense mutation in *AARS2*, which encodes the mitochondrial alanyl-tRNA synthetase (mtAlaRS). There is a possibility that tRNA editing may lead to different protein repertoire which may disrupt the function of different enzymes causing cardiomyopathy.

Micro RNA

miRNAs are the non-coding, highly conserved RNA sequences found in intronic and intergenic regions of the genome and in gene expression by binding to the 3'untranslated region (3'UTR) of messenger RNAs (mRNAs), resulting in repression of translation. In cardiomyocytes or cardiac cells, micro RNAs act as important regulators of gene expression in many biological processes.

In the human genome, about 1000 miRNAs exist. Recent studies have indicated that target recognition by one miRNA is mainly achieved by pairing to the nucleotides at positions 2–7 of the miRNA (the so-called "miRNA seed"), and often a single mutation in the miRNA seed match of the target site can lead to total abolishment of effective repression [30]. miRNAs are transcribed by RNA polymerase II into long primary miRNA (pri-miRNA) transcripts which are capped and polyadenylated.

Role of 'edited' miRNAs in cardiac diseases

Increasing expression of miR-1 and miR-133 has been found in neonatal hearts and substantially higher levels are maintained in adult cardiac tissues, suggesting that in addition to regulating myogenesis, they may also possess other cellular functions in adult cardiac cells. A study was carried out to localize editing sites in miRNAs, playing a role in cardiac diseases. It clearly showed an occurrence of A to I editing with both peaks of A and G shown at equivalent positions correlating to RNA editing.
Circulating miRNAs, like miR-208, miR-133, miR-194 and miR-1, are novel biomarkers in AMI diagnosis and prognosis with high sensitivity and specificity. Circulating miRNAs are endogenous miRNAs that are detectable in whole blood, serum or plasma in a highly stable form.

a) miR-1: miR-1, like miR-133, is highly expressed in cardiac and skeletal muscle cells. It has been shown that miRNA plays an important role in the development of the embryonic heart. The expression of ADAR1 is down regulated by miR-1. Moreover, the earliest and highest expression of ADAR1 is detected in a mouse's embryonic heart at E10.0, suggesting an important function of ADAR1 in mouse heart development and its tight regulation by miR-1 in the KCNE1 gene is very important in LQT5 syndrome.

b) miR-133: miR-133 is found to be down regulated in cardiac hypertrophy condition and is enriched in cardiac and skeletal muscle cells. This miRNA is present in the genome as two copies (miR-133a-1 and miR-133a-2). Copy miR-133a-2 is hosted within a gene transcribed in the same direction as the annotated miRNA gene. In contrast, copy miR-133a-1 overlaps a gene transcribed from the opposite strand. Antisense miRNA miR-133a-1 was shown to undergo A to I editing. Loss-of-function mutations in KCNQ1 causes Long QT syndrome type I (LQT1) [31]. But detailed analysis of the 3'UTR of KCNQ1 showed that there are four putative binding sites for miR-133 based solely on complementarity, each of them containing six nucleotides exactly matching the 2–7 nucleotides from the 5'-end of miR-133. Similarly miR-133 is found to regulate the levels of CTGF by targeting the CTGF 3'UTR which in turn contributes to the development of cardiac fibrosis as in ARVD/C, LQT and cardiomyopathy.

c) miR-151: miR-151 belongs to a class of miRNA that is derived from genomic repetitive sequences, LINE2. It has shown to undergo A to I editing but at relatively low levels in heart. The editing site, in case of miR-151, was found at a position 2 to 7 bases from the 5' end of the miRNA. Although RNA editing of miR-151 has been observed in the heart, it is unknown whether it plays a role in cardiac diseases. RNA editing of pri-miR-151 at two specific positions within its fold back dsRNA structure completely inhibits its cleavage by the Dicer–TRBP complex, leading to an accumulation of edited pre-miR-151 intermediate RNA which has an important role in miRNA biogenesis.

d) miR-194: miR-194 is reported to be p53-responsive. miR-151, miR-34a and miR-192 were shown to act as predictive indicators of heart failure after acute myocardial infarction and function as circulating regulators of heart failure development via the p53 pathway.

e) miR-208: Some miRNAs (miR-1, miR-133, miR-206 and miR-208) are muscle specific hence called myomiRs. miR-208 is encoded in an intron of the alpha-MHC gene (MYH-7). The miR-208 family consists of two transcripts, miR-208a and miR-208b. miR-208a is encoded by an intron of its host gene, alpha-MHC, while miR-208b is encoded by beta-MHC. Both miR-208a and alpha-MHC are cardiac specific and are concurrently expressed during development, suggesting that their expression levels are controlled by a common regulatory element. miR-208a is shown to be involved in regulating MHC isoform switching. Among the strongest predicted targets of miR-208 is the thyroid hormone receptor associated protein 1 (THRAP1), which can exert positive and negative effects on transcription. THRAP1 acts as a co-regulator of the thyroid receptor to suppress β-MHC expression.

miR-208 was shown to pair with a specific RNA region of Med13 and inhibit its translation. This in turn reduces thyroid receptor signalling, leading to increased β-MHC expression and predictably poorer cardiac function. RNA editing was observed in the hyper edited region of the *MED13* gene (THRAP1) within an Alu element in intron 9. Editing occurs in clusters and proved to undergo extensive RNA editing in 27 putative A-to-I RNA editing sites. The degree of RNA editing in this region may serve as a general reflection of the extent of RNA editing taking place in the transcriptome of cyanotic CHD patients and as a possible mechanism of epigenetic regulation underlying cyanotic CHD pathophysiology [32].

CONCLUSION

RNA editing is a site-specific modification which frequently leads to protein diversity. APOBEC2, due to its highly tissue-specific distribution and highly conserved nature, may have a specific and unique function in muscle and heart tissue although its exact role in the heart is still obscure. Alu repeat elements are the most enriched sites for RNA editing and the two inverted Alu repeats forming dsRNA may disrupt the function of various genes, especially *MYBPC3*. RNA editing events may contribute to channelopathies like LQT syndrome which are yet to be determined. Hence the occurrence of RNA editing events remains elusive, nevertheless it is important in the evolution perspective. The role of RNA editing in miRNAs is established, hence it can be considered as potential biomarkers in a wide range of diseases. RNA editing of miR-133, miR-208 and miR-151 have shown to up/down regulate these mi-RNAs which in turn leads to poorer cardiac function. Loss of miR-133 in hypertrophy and heart failure suggests that an increased expression of miR-133 is reported. The possible RNA editing role in disorder related genes can open new avenues in the development of novel therapeutics.

REFERENCES

[1] Gott, Jonatha M., and Ronald B. Emeson. "Functions and mechanisms of RNA editing." Annual review of genetics 34.1 (2000): 499-531.
[2] Schaub, Myriam, and Walter Keller. "RNA editing by adenosine deaminases generates RNA and protein diversity." Biochimie 84.8 (2002): 791-803.
[3] Maas, Stefan, Alexander Rich, and Kazuko Nishikura. "A-to-I RNA editing: recent news and residual mysteries." Journal of Biological Chemistry 278.3 (2003): 1391-1394.
[4] Jepson JE, Reenan RA. RNA editing in regulating gene expression in the brain. Biochim Biophys Acta 1779(8):459-470, 2008.
[5] Wedekind, Joseph E., et al. "Messenger RNA editing in mammals: new members of the APOBEC family seeking roles in the family business." Trends in Genetics 19.4 (2003): 207-216.
[6] Muramatsu, Masamichi, et al. "Class switch recombination and hypermutation require activation-induced cytidine deaminase (AID), a potential RNA editing enzyme." Cell 102.5 (2000): 553-563.
[7] Hundley, Heather A., and Brenda L. Bass. "ADAR editing in double-stranded UTRs and other non-coding RNA sequences." Trends in biochemical sciences 35.7 (2010): 377.

[8] Nishikura, Kazuko. "Functions and regulation of RNA editing by ADAR deaminases." Annual review of biochemistry 79 (2010): 321.

[9] Keegan, Liam P., et al. "Adenosine deaminases acting on RNA (ADARs): RNA-editing enzymes." Genome Biol 5.2 (2004): 209.

[10] Keller, Walter, Jeannette Wolf, and André Gerber. "Editing of messenger RNA precursors and of tRNAs by adenosine to inosine conversion." FEBS letters 452.1 (1999): 71-76.

[11] Jepson, James EC, and Robert A. Reenan. "RNA editing in regulating gene expression in the brain." Biochimica et Biophysica Acta (BBA)-Gene Regulatory Mechanisms 1779.8 (2008): 459-470.

[12] Cenci, Caterina, "Down-regulation of RNA editing in pediatric astrocytomas ADAR2 editing activity inhibits cell migration and proliferation." Journal of Biological Chemistry 283.11 (2008): 7251-7260.

[13] Tang, Wei, Yongjun Fei, and Michael Page. "Biological significance of RNA editing in cells." Molecular biotechnology 52.1 (2012): 91-100.

[14] Feng, Yi, "Altered RNA editing in mice lacking ADAR2 autoregulation." Molecular and cellular biology 26.2 (2006): 480-488.

[15] Tonkin, Leath A., "RNA editing by ADARs is important for normal behavior in Caenorhabditis elegans." The EMBO journal 21.22 (2002): 6025-6035.

[16] Palladino, Michael J., "A-to-I Pre-mRNA Editing in< i> Drosophila</i> Is Primarily Involved in Adult Nervous System Function and Integrity." Cell 102.4 (2000): 437-449.

[17] Chen, C. X., "A third member of the RNA-specific adenosine deaminase gene family, ADAR3, contains both single- and double-stranded RNA binding domains." Rna 6.5 (2000): 755-767.

[18] Melcher, Thorsten, "RED2, a brain-specific member of the RNA-specific adenosine deaminase family." Journal of Biological Chemistry 271.50 (1996): 31795-31798.

[19] Wedekind, Joseph E., "Messenger RNA editing in mammals: new members of the APOBEC family seeking roles in the family business." Trends in Genetics 19.4 (2003): 207-216.
[20] Liao W, Hong SH, Chan BH, Rudolph FB, Clark SC, ChanL. "APOBEC-2, a cardiac- and skeletal muscle-specific member of the cytidine deaminase supergene family". Biochem Biophys Res Commun , 1999 Jul 5;260(2):398-404.
[21] Gray, Michael W. "Evolutionary origin of RNA editing." Biochemistry 51.26 (2012): 5235-5242.
[22] Kim, D.D., Kim, T.T., Walsh, T., Kobayashi, Y., Matise, T.C., Buyske, S., and Gabriel, A. 2004. Widespread RNA editing of embedded Alu elements in the human transcriptome Genome Res. 14: 1719-1725.
[23] Levanon, E.Y., Eisenberg, E., Yelin, R., Nemzer, S., Hallegger, M., Shemesh, R., Fligelman, Z.Y., Shoshan, A., Pollock, S.R., Sztybel, D., et al. 2004. Systematic identification of abundant A-to-I editing sites in the human transcriptome. Nat. Biotechnol. 22: 1001-1005.
[24] Blow, Matthew, "A survey of RNA editing in human brain." Genome research 14.12 (2004): 2379-2387.
[25] Pezzoli, Laura, "A new mutational mechanism for hypertrophic cardiomyopathy." Gene 507.2 (2012): 165-169.
[26] Towbin, J. A., and N. E. Bowles. "The failing heart." Nature 415.6868 (2002): 227-233.
[27] McCartan, Charles, "Cardiomyopathy Classification: Ongoing Debate in the Genomics Era." Biochemistry research international 2012 (2012).
[28] Bhalla T, Rosenthal JJ, Holmgren M, Reenan R: Control of human potassium channel inactivation by editing of a small mRNA hairpin. Nat Struct Mol Biol 2004, 11:950-956.
[29] Tan, Bao Zhen, "Dynamic regulation of RNA editing of ion channels and receptors in the mammalian nervous system." Mol Brain 2.1 (2009): 13.
[30] Lewis, Benjamin P., Christopher B. Burge, and David P. Bartel. "Conserved seed pairing, often flanked by adenosines, indicates that thousands of human genes are microRNA targets." cell 120.1 (2005): 15-20.

[31] Catalucci, D., M. V. G. Latronico, and G. Condorelli. "MicroRNAs and the Control of Heart Pathophysiology." RNA Technologies in Cardiovascular Medicine and Research. Springer Berlin Heidelberg, 2008. 53-68.
[32] Borik, Sharon, "Increased RNA editing in children with cyanotic congenital heart disease." Intensive care medicine 37.10 (2011): 1664-1671.

CHAPTER SIX

Reduction of Vascular Dysfunction and Morbidity in Patients with Community Acquired Pneumonia by Using Dexamethasone

Laith Abdulhussein Aldabbagh[1]

[1] *Internal medicine department, Kufa medical college, Najaf, Iraq*

Email Address: laitha.aldabbagh@uokufa.edu.iq

Abstract

Organ dysfunction can accompany systemic inflammatory responses which occur in patients with community acquired pneumonia (CAP). The aim of this study is to determine whether dexamethasone improves vascular dysfunction and morbidity in patients with this disorder who are over 40 years of age. A total of 99 patients with the diagnosis of CAP were divided randomly into two groups. Group 1, which comprises 58 patients, received the standard therapy for CAP in addition to intravenous dexamethasone. Group 2, consisting of 41 patients, received only the standard therapy for CAP. All patients were followed up on the first and fourth day of their hospital admission and then assessed 30 days later. The mean age of the patients is 56 years. Of the 99 patients, 49 were males and 50 were females. There was a significant improvement in the group that was administered the dexamethasone. From day 1 to day 4 post admission there was a notable recovery in the total arterial compliance and arterial impedance in group 1 in comparison with group 2.

Keywords: Dexamethasone, Community acquired pneumonia (CAP), Vascular dysfunction.

INTRODUCTION

Community acquired pneumonia (CAP) is an important contributor for mortality and morbidity all over the globe. Statistics state that 20% of all attacks of pneumonia result in hospital admissions.

Elderly patients may be at risk for severe infections with high morbidity and mortality rates [1].

Impairment of the function of many organs can accompany the systemic inflammatory reaction which occurs in patients with CAP. The addition of corticosteroid therapy to the standard management of CAP might mitigate the intensity of these reactions [2]. Corticosteroids are an important suppressor of inflammation. Therapies with this group of steroids reduce pro-inflammatory cytokine transcription and hence reduce systemic and pulmonary inflammation in CAP [3, 4].

CAP patients have an elevated level of endothelin-1, an endothelium-specific vasoconstriction peptide [5]. Blood concentration of endothelin-1 at the time of admission to the hospital have a significant correlation with the severity and the prognosis of the pneumonia [6].

A decrease in peripheral vascular resistance occurs during the acute phase of pneumonia in young and middle-aged people [7]. However, increased peripheral vascular resistance that is only partly responsive to volume expansion can occur in up to a third of middle-aged patients [8]. This finding suggests that vasoactive responses leading to increased cardiac afterload (i.e. increased peripheral vascular resistance) can also be present in elderly individuals with pneumonia, which is potentially counterproductive in patients with chronic impairment of myocardial function [9].

METHODOLOGY

In this case controlled study, 99 patients with a diagnosis of community acquired pneumonia were admitted to the teaching hospital from June 2012 to March 2014.

Inclusion Criteria:

- Patients aged 40 to 80 years with community acquired pneumonia.

- Criteria used for diagnosis of community acquired pneumonia.

Chest radiograph findings consistent with new pulmonary opacities in addition to any two of the following features:

1. Cough.
2. Production of sputum.
3. Temp >38.0 °C or <36.0 °C.
4. Audible abnormalities by chest examination compatible with pneumonia.
5. Leukocytosis (>10.000 cells/mm3) or leukopenia (<4000 cells/mm3).
6. Positive CRP blood test.

Exclusion Criteria:

- Patients who have received chemotherapy, immunosuppressive or immune-modulator medications (cyclosporine, mycophenolate mofetil, azathioprine).
- Patients who have taken corticosteroids in the past 12 weeks.

- Patient with a history of diabetes mellitus.
- Patients with a history of hypertension or are on antihypertensive drugs.
- Patients with chronic obstructive pulmonary disease who are on steroids.
- Patients who require respiratory care unit treatment.
- Patients with a history of overt cardiovascular disease.

A comprehensive explanation of the purpose and design of the study, which was conducted according to the ethical principles stated in the Declaration of Helsinki, was approved by the Kufa university medical ethics committee. Written informed consents were obtained.

All patients had been subjected to a full history inquiry and complete physical examination along with assessment of the pneumonia severity according to CURB 65 score, a pneumonia severity score consisting of 5 parameters that assigns one point for every positive parameter. The first parameter is the estimation of the degree of confusion by an abbreviated mental test score. If the test score is < = 8 then one point was added. Blood urea nitrogen values more than 19 mg/dL add one point. Respiratory rate assessment that exceeds 30 per minute sums up one point. Blood pressure estimation with diastolic blood pressure less than 60 mmHg or with systolic blood pressure less than 90 mmHg totals one point. If the age of the patient is > = 65 one point was added. In the end the total score was calculated. Results that are 0-1 points indicate low risk while a total score of 2-5 points indicate higher risk [10]. Patients were sent for a chest x ray, for a complete blood cell count (4ml samples were drawn), blood urea and CRP. Next all patients were sent for assessment of total arterial compliance and arterial impedance.

The patients were divided randomly into two groups. The first group (n= 58) were allocated to receive a single dose, at 6 AM, of 4 mg of intravenous dexamethasone daily for a period of 3 days in addition to the standard therapy for CAP (according to the IDSA/ATS guidelines for CAP in adults [11]). The second group (n= 41) were designated to receive the standard therapy for CAP exclusively.

All patients had been followed up on the first and fourth day of their hospital admission and then assessed 30 days later. The patients were followed up with and monitored for the development of hyperglycemia, gastrointestinal upset and pneumonia complications such as pleural effusion and the need for mechanical ventilation. Thirty days readmission need, mean time of switch to oral therapy in days, the mean of the length of hospital stay in days and the improvement in vascular dysfunction (which is assessed by echocardiography by examining total arterial compliance as measured by calculating the stroke volume index pulse-pressure ratio [SVi/PP] and arterial impedance as measured by the valvulo-arterial impedance [Zva]) all calculated by the formula (systolic blood pressure + mean gradient over the aortic valve)/ stroke volume index.

SVi/PP >0.6Ml/m2/mmHg, and Zva$=<4$mmHg/mL/m2 were considered normal (12).

All statistical analyses were done using SPSS version 20 software (SPSS Inc. Chicago, IL, USA). Categorical data were compared using chi-square tests. Student t tests were used to compare numerical variables. The data was expressed as a mean (±standard deviation; SD) and the numbers where appropriate. P value < 0.05 was considered statistically significant.

Table 1 shows the patients' characteristics of the two groups.

variable	First group	Second group	P value
Mean age - years	57.3	56.6	0.5
Female – number	28	22	0.6
Smoker- number	25	15	0.7
Mean of the body mass index –kg/m^2	23.5	24.4	0.8
Confusion (number of patient with abbreviated mental score less than or equal to 8)	27	25	0.1
Blood urea nitrogen (number of patients with blood urea nitrogen more than 19 mg/dL.)	13	8	0.7
Raised respiratory rate (number of patients with respiratory rate more than 30 per minute)	16	14	0.4
Blood pressure: (number of patients with diastolic blood pressure of less than 60 mmHg or systolic blood pressure of less than 90 mmHg)	17	15	0.4
Age > = 65 years (number of patients with age > = 65 years)	31	23	0.7
Total CURB 65 score 2-3(number of patients)	43	31	0.8
Mean of total cholesterol (mg/dL)	116	107	0.4
Mean of triglyceride – (mg/dL)	132	118	0.3
Mean of HDL- (mg/dL)	53	81	0.02
Mean of total arterial compliance, SVi/PP (Ml/m2/mmHg)	0.5	0.5	0.1
Mean of arterial impedance, Zva (mmHg/mL/m2)	5	4.9	0.09
Total number of patients	58	41	0.08

RESULTS AND DISCUSSION

The patients' ages range from 40 to 80 years and the mean age was 56 (±9.4) years. Of the 99 patients analyzed, 49 patients were male and 50 were female. All variables were similar between the two groups as shown in "table 1".

Table 2: shows the difference in the study outcomes among the two groups.

Patient groups	First group(58)	Second group(41)	P value
Development of hyperglycemia	8	1	0.05
Development of gastrointestinal upset	1	1	0.8
Development of pneumonia complications (like pleural effusion)	1	2	0.3
Need for mechanical ventilation	0	3	0.03
30 days readmission need	0	4	0.01
Mean time of switch to oral therapy in days	4	6	<0.002
Mean of mean length of hospital stay in days	4	6	<0.002

There was a considerably enhanced outcome (need for mechanical ventilation, 30 days readmission need, mean time of switch to oral therapy in days, the mean of length of hospital stay in days and the improvement in vascular dysfunction) in the group that dexamethasone was administered to (in addition to the standard management of CAP) as shown in "table 2".

A large number of patients with CAP have vascular dysfunction manifested by abnormalities in the systemic arterial compliance and systemic arterial impedance early on during the course of the disease that were corrected to a compelling degree in the group that administered dexamethasone in addition to the standard management of CAP as shown in " table 3,4,5 and 6".

Table 3: shows the association between total arterial compliance at day 1 and day 4 following hospital admissions among the two groups.

Patients groups	Abnormal total arterial compliance Day 1	Abnormal total arterial compliance Day 4	P value
First group(n=58)	20	5	0.0007
Second group(n=41)	11	7	0.2

Table 4: shows the association between arterial impedance at day 1 and day 4 following hospital admissions among the two groups.

Patients groups	Abnormal arterial impedance Day 1	Abnormal arterial impedance Day 4	P value

First group(n=58)	21	9	0.01
Second group(n=41)	15	9	0.1

Table 5: shows the association between the mean of total arterial compliance at day 1 and day 30 following hospital admissions, among the two groups.

group	Mean of total arterial compliance, SVi/PP (Ml/m2/mmHg) at day 1	Mean of total arterial compliance, SVi/PP (Ml/m2/mmHg) at day 30	P value
First group	0.5	0.6	0.0001
Second group	0.5	0.5	0.45

Table 6: shows the association between the mean of arterial impedance at day 1 and day 30 following hospital admissions among the two groups.

group	Mean of arterial impedance, Zva (mmHg/mL/m2) at day 1	Mean of arterial impedance, Zva (mmHg/mL/m2) at day 30	P value
First group	5	3.2	0.001
Second group	4.9	4.5	0.47

Pneumonia has significant effects on the cardiovascular system regardless of the degree of severity (in spite of being by definition confined to the lungs [13, 14].)

According to Corrales-Medina, the risk for cardiovascular complications is higher in the first few days after a pneumonia diagnosis, with about 90% of these events recognized within 7 days of diagnosis. It should be noted that more than half of these events are identified within the first 24 hours [15, 16].

In correlation with previous observations, this study showed that a significant number of patients with CAP have vascular dysfunction manifested by abnormalities in the systemic arterial compliance and systemic arterial impedance early on during the course of the disease.
According to Benson H et al. and Kumar R et al, increased peripheral vascular resistance can occur in up to a third of middle aged patients [8, 17], since patients with pneumonia have transiently raised serum concentrations of endothelin-1 (an endothelium-specific vasoconstriction peptide [18].) In addition, vasoactive responses leading to increased cardiac afterload (i.e. increased peripheral vascular resistance) can also be present in elderly individuals with pneumonia, which is potentially counterproductive in patients with chronic impairment of myocardial function [19]. In the present study, dexamethasone, in addition to standard management of CAP, can reduce vascular dysfunction manifested as abnormal systemic arterial compliance and impedance which are an indication of arterial stiffness and endothelial dysfunction that occurs in those patients. This may explain in part the improved morbidity and mortality outcomes in patients treated with dexamethasone. This finding may highlight the role of CAP as a trigger for adverse cardiovascular outcome in the future, especially if dexamethasone and other

measures to reduce vascular dysfunction are not implemented energetically.

Evidence has been accumulating for several years that protecting the endothelium is the key to reduce cardiovascular risk (CV). Endothelial dysfunction results in reduced compliance or increased arterial stiffness, particularly in the smaller arteries. This abnormality is a characteristic of patients with hypertension but may also be seen in normotensive patients before the appearance of clinical disease. Reduced arterial compliance is also seen in patients with diabetes and in smokers, and is part of a vicious cycle that further elevates blood pressure, aggravates atherosclerosis, and leads to increased CV risk. Although other factors are involved, the damage to the endothelium results in reduced secretion of nitric oxide, which influences smooth muscle growth, migration, and contraction, as well as influencing inflammation and clotting [20].

According to Pierre-Francois Laterre, it is increasingly accepted that severe CAP is a condition in which abnormal coagulation and an excessive host inflammatory response play an important role. Based on this understanding, a number of strategies for reducing morbidity and mortality are being pursued. Among them is the use of glucocorticoid or other anti-inflammatory agents. There appears to be a correlation between persistent inflammation (evident based on raised levels of IL-6, for example) and cardiovascular events occurring even during the weeks and months after pneumonia. This might justify further investigation into an anti-inflammatory strategy [21].

CONCLUSION

It is concluded that Dexamethasone reduces the need for mechanical ventilation, the necessity of the 30 day readmission, the mean time of switching to oral therapy and the mean of the length of hospital stay.

Vascular dysfunction manifested as abnormalities in systemic arterial compliance and impedance can accompany a significant number of cases with CAP. Dexamethasone, in addition to the standard management of CAP, reduces vascular dysfunction that accompanies the disease.

REFERENCES

[1] F.Alvares-lerma, A. Torres, Sever community –acquired pneumonia, Curr. Opin. Crit. Care. 2004; 10.
[2] Meijis S, Hardeman H, Remmelts H, Dexamethasone and length of hospital stay in patient with community acquired pneumonia: a randomized, double-blind, placebo-controlled trial. Lancet. 2011; 377: 2023-2030.
[3] Monton C, EwigS Torres A, Role of glucocorticoids on inflammatory response in nonimmunosuppressed patients with pneumonia: a pilot study. EurRespir J. 1999; 14:218–220.
[4] Rosewicz S, McDonald AR, Maddux BA, Mechanism of glucocorticoid receptor down-regulation by glucocorticoids. J BioChem. 1988; 263:2581–2584.
[5] Schuetz P, Christ-Crain M, Zimmerli W, Mueller B. Repeated measurements of endothelin-1 precursor peptides predicts the outcome in community-acquired pneumonia. Intensive Care Med 2011; 37: 970–80.
[6] Schuetz P, Wolbers M, Christ-Crain M, Pro-hormones for prediction of adverse medical outcome in community-acquired

pneumonia and lower respiratory tract infections. Crit Care 2010; 14: R106.

[7] Kumar R, Wallace WA, Ramirez A, Benson H, Abelmann WH. Hemodynamic effects of pneumonia. II. Expansion of plasma volume. J Clin Invest 1970; 49: 799–805.

[8] Benson H, Akbarian M, Adler LN, Abelmann WH. Hemodynamic effects of pneumonia. I. Normal and hypodynamic responses. J Clin Invest 1970; 49: 791–98.

[9] Roger VL, Go AS, Lloyd-Jones DM, Heart disease and stroke statistics—2011 update: a report from the American Heart Association. Circulation 2011; 123: e18–209.

[10] Lim WS, van der Eerden MM, Laing R, Defining community acquired pneumonia severity on presentation to the hospital: an international derivation and validation study. Thorax. 2003 May;58(5):377-82.

[11] Lionel A. Mandell, Richard G. Wunderink, Infectious Diseases Society of America/American Thoracic Society Guidelines on the Management of Community Acquired Pneumonia in Adults: Clinical Infectious Diseases. 2007; 44:S27.

[12] Briand M, Dumesnil JG, Kadem L, Tongue AG, Rieu R, Garcia D, Pibarot P. Reduced systemic arterial compliance impacts significantly on left ventricular afterload and function in aortic stenosis: implications for diagnosis and treatment. J Am CollCardiol. 2005; 46:291–298.

[13] Khand AU, Gemmell I, Rankin AC, Cleland JG. Clinical events leading to the progression of heart failure: insights from a national database of hospital discharges. Eur Heart J 2001; 22: 153–64.

[14] Corrales-Medina VF, Madjid M, Musher DM. Role of acute infection in triggering acute coronary syndromes. Lancet Infect Dis 2010; 10: 83–92.).

[15] Corrales-Medina VF, Musher DM, Wells GA, Chirinos JA, Chen L, Fine MJ. Cardiac complications in patients with

community-acquired pneumonia: incidence, timing, risk factors, and association with short-term mortality. Circulation 2012; 125: 773–81.
[16] Corrales-Medina VF, Serpa J, Rueda AM, Acute bacterial pneumonia is associated with the occurrence of acute coronary syndromes. Medicine (Baltimore) 2009; 88: 154–59.
[17] Kumar R, Wallace WA, Ramirez A, Benson H, Abelmann WH. Hemodynamic effects of pneumonia. II. Expansion of plasma volume. J Clin Invest 1970; 49: 799–805.
[18] Schuetz P, Christ-Crain M, Zimmerli W, Mueller B. Repeated measurements of endothelin-1 precursor peptides predicts the outcome in community-acquired pneumonia. Intensive Care Med 2011; 37: 970–80.
[19] Roger VL, Go AS, Lloyd-Jones DM, Heart disease and stroke statistics—2011 update: a report from the American Heart Association. Circulation 2011; 123: e18–209.
[20] Jay N. Cohn, Arterial compliance to stratify cardiovascular risk: more precision in therapeutic decision making, Am J Hypertens (2001) 14 (S5): 258S-263S doi: 10.1016/S0895 7061(01)02154-9.
[21] Pierre-Francois Laterre, Beyond antibiotics in severe community-acquired pneumonia: the role and rationale for tissue factor pathway inhibition, Critical Care 2008, 12, 26 November 2008 (Suppl 6):S4 (doi: 10.1186/cc7027).

CHAPTER SEVEN

EPIDEMIOLOGICAL STUDY OF SHIGELLA SPP. IN AL-MIQDADIYAH CITY AND ITS SENSITIVITY FOR ANTIBIOTICS

Z. N. Al-karakooly[1,*] and Q. A. Al-anbaky[1, 2], S. M. Al-dahalakhi[1], T. Aljasem[3]

[1] *College of Science University of Diyala, Baquba, Diyala, Iraq*

[2]*Department of Biology, College of Science, Art and Letters, University of Arkansas at Little Rock, Little Rock, AR 72204;* [3] *Al Zahra hospitals, AL- Miqdadiyah, Diyala, Iraq*

Correspondence: znalkarakoo@ualr.edu

ABSTRACT

This study focuses on the isolation and diagnosis of 30 isolates of the bacterium Shigella Spp collected from 180 stool specimens and rectal swabs of patients (children and adults). Samples were taken from subjects with watery and bloody diarrhea in AL- Miqdadiyah General Hospital and Al Zahra'a hospital. All samples were collected and grown on among different agricultural media to investigate the presence of the Shigella and all the isolates were subjected to biochemical tests. 30 isolates of Shigella spp were being investigated on to determine their sensitivity towards 10 antibiotics which were: Amoxicillin, Ampicillin, Cephataxim, Chloramphenicol, Ciproflaxacin, Cotrimoxazol, Nalidixic acid, Neomycin, Rifampicin and Tetracyclin. Isolates showed 100% resistance for Cephataxim, Ciproflaxacin, Cotrimoxazol and Rifampicin. On the other hand, sensitivity reached 100 % for Neomycin and Nalidixic acid. Some isolates were fluctuating in resistance and sensitivity to Ampicillin, Amoxicillin, Chloramphenicol and

Tetracyclin.

KEYWORDS: Shigella, Epidemiology, Sensitivity, Antibiotics.

INTRODUCTION

Bacillary dysentery disease is one of the most prevalent diseases in poor regions of the world as well as industrial and developing countries. It is one of the endemic diseases in those areas [1]. The most important factors that help to spread diseases are rapidly declining living conditions, lack of health awareness, malnutrition and living in crowded and confined spaces [1, 2]. A small dose of Shigella (10- 100) cells is sufficient to cause cell injury [1, 2, 3].

Shigella can grow in food and milk causing a change in appearance and/or taste. They have the ability to survive for up to six months in tap water and ice [4]. The course of the infection, bacillary dysentery, ranges from six months to ten years. The occurrence of the disease takes place due to living conditions such as malnutrition. Such conditions may be meet when travelling to poor areas of the world [5, 6, 7, 8].

Incidences of infection by Shigella is more reluctant in the hot and humid seasons compared to the cold seasons and there is no difference in incidence of infection between males and females. The incidence of bacillary dysentery in subjects less than six months of age is rare since breastfeeding plays an important role in its reduction. Mother's milk contains antibodies like IgA type against LPS (Lipopolysaccharids) and plasmid antigens for virulence of the bacterium Shigella [7, 9].

One route of transmission of Shigella is from feces of an infected person to the mouth of another. Food and water contaminated with bacteria by flies is another. Sexual contact by the anus is also an important way in the transmission of the disease [2, 3, 10, 11].

Incubation period for this bacterium in the human body ranges between one to seven days and generally 48 hours is sufficient for the breeding of the bacteria and the emergence of symptoms of the disease [2]. The frequency of infection varies from one country to another and it has been found that the type Shigella sonnei was the causative agent of the disease bacillary dysentery in developed countries and industrial zones, while Shigella flexneri comes second [11]. However in both developing countries and poor areas Shigella flexneri occupies the first place and Shigella sonni in second place. The type Shigella boydii is rare and is responsible for diarrhea or bacillary dysentery [9, 12]. The infections occurring due to Shigella dysenteriae make up for the number one form of endemic diseases in Asia and in developed countries [1, 3, 9]. Estimated annual rate of the infection bacillary dysentery caused by Shigella is 100-150 million infections. [5, 8, 11], while the number of deaths is estimated at about 1.1 million. [6, 8, 10, 13].

The bacterium Shigella is one of the most important causes of diarrhea, especially in children under five years of age. There are a few studies about its role in the incidence of diarrhea in the Diyala province. This study has focused on isolation and diagnosis of Shigella from feces of patients with acute watery and bloody diarrhea (children and adults) by using chemical and biochemical methods as well as an examination of the sensitivity of the bacteria towards some common antibiotics.

MATREIALS AND METHODS

Collection Samples

180 samples of stool and rectal swabs were collected from patients (children and adults) suffering from acute watery and bloody diarrhea in AL- Miqdadiyah General Hospital and Al Zahra Hospital for delivery and children in the town of AL-Miqdadiyah. We have investigated the bacteria Shigella in those

samples that were collected from 40 adults and 140 children under five years of age. After collecting samples using a sterile swab, they were planted during a period of half an hour on a specific culture media like MacConky agar or water peptone and then incubated at a temperature °C 37 for 24 hours.

DIAGNOSIS

Cultural Characteristics

After the appearance of colonies grown on the center of the agar MacConky, the bacteria were diagnosed depending on the size, color, and height.

Gram Stain

Bacterial smears were done from non-fermented sugar lactose colonies of bacteria that were grown on the MacConkey agar. The bacterium were diagnosed initially depending on some characteristics, which include formal colony size, color, shape, height and the studied characteristics of the constituent cells of these colonies under the microscope after staining with grams stain.

Biochemical Tests

Diagnosis of biochemical tests was done depending on the methods described by [1, 14, 15]. These tests include the indole production test using the peptone water as a cultural medium, fermentation of sugar mannitol test using mannitol agar semi-solid, fermentation sugar glucose test, lactose test and the production of CO2 gas, Oxides test and catalase test production by using a 3% solution of H_2O_2 solution.

Antibiotic Sensitivity Tests

Bauer et. al. came up with a method in 1966 [16] that was used to test the sensitivity of bacterial isolates towards ten types of antibiotics like Amoxicillin (Am), Ampicillin (Amp), Cephataxim (CTX), Chloramphenicol (C), Ciproflaxacin (Cip), (SXT) Cotrimoxazol, (Na) Nalidixic acid, (N) Neomycin, Rifampicin (Rif), Tetracyclin (TE)

RESULTS AND DISCUSSION

Isolation and Diagnosis of the Shigella Spp.

Microscopic and Cultural Characteristics

Shegilla colonies on the MacConkey agar are small, smooth, round and pale yellow in color due to lack of ability to ferment lactose. Microscopically, they were gram-negative, rod-meningococcal and single cells.

Biochemical Tests

Table 2 shows the results of basic biochemical tests for 30 isolates of the genus Shegilla. The results of these tests were adopted to distinguish and differentiate among the bacteria Shigella and other species belonging to the family of intestinal and private races such as Salmonella, Escherichia coli, and the genus Proteus. Susceptibility of the bacterial isolates on the fermentation of glucose was studied as well as gas production of CO_2, the fermentation of lactose, mannitol fermentation of sugar and production of an indole ring. Common to all isolates of the bacterium Shigella in usability on the fermentation of sugar glucose and non-production of CO2 gas from the fermentation of this sugar in addition to its inability to ferment on lactose with an exception to isolates Shigella Spp28, 29, 30 marked by slow fermentation for this sugar which is called "Late lactose ferment". In the production of a ring indole, consumption of the amino acid tryptophan and fermented sugar mannitol can be divided into two groups, fermented and non-fermented sugar. [1, 14, 15].

The results were positive for the Catalase exam. Shigella Spp1 was negative for this examination. Checking oxides showed a negative result for all isolates and this means that the Shigella do not have the enzyme oxides.

Table 1: Shows the age distribution of the patients surveyed.

Age groups	the number of patients who suffer from diarrhea cases	the number of infections with the bacterium Shigella
less than years	38	3
more than 1-4 year	70	14
more than5-9 years	22	3
more than 10-14 years old	10	3
more than 15 years and older (adults)	40	7

Table 2: Biochemical tests, basic differentials of the genus Shigella.

Isolates #	Type	Glu.	Lac.	CO_2 Pro.	Mann.	Indol Pro.	Cata test	Oxi test
1	Shigella Spp.	+	-	-	-	-	-	-
2	Shigella Spp	+	-	-	-	+	+	-
3	Shigella Spp	+	-	-	-	+	+	-
4	Shigella Spp	+	-	-	-	+	+	-
5	Shigella Spp	+	-	-	+	+	+	-
6	Shigella Spp	+	-	-	+	-	+	-
7	Shigella Spp	+	-	-	+	-	|	-
8	Shigella Spp	+	-	-	+	+	+	-
9	Shigella Spp	+	-	-	+	+	+	-
10	Shigella Spp	+	-	-	+	+	+	-
11	Shigella Spp	+	-	-	-	-	+	-
12	Shigella Spp	+	-	-	-	-	+	-
13	Shigella Spp	+	-	-	-	-	+	-
14	Shigella Spp	+	-	-	-	-	+	-
15	Shigella Spp	+	-	-	-	-	+	-
16	Shigella Spp	+	-	-	-	-	+	-
17	Shigella Spp	+	-	-	-	-	+	-

18	Shigella Spp	+	-	-	-	-	+	-
19	Shigella Spp	+	-	-	-	-	+	-
20	Shigella Spp	+	-	-	-	-	+	-
21	Shigella Spp	+	-	-	+	-	+	-
22	Shigella Spp	+	-	-	+	-	+	-
23	Shigella Spp	+	-	-	+	-	+	-
24	Shigella Spp	+	-	-	+	-	+	-
25	Shigella Spp	+	-	-	+	-	+	-
26	Shigella Spp	+	-	-	-	-	+	-
27	Shigella Spp	+	-	-	-	-	+	-
28	Shigella Spp	+	*+	-	-	+	+	-
29	Shigella Spp	+	*+	-	+	-	+	-
30	Shigella Spp	+	*+	-	+	-	+	-

Table 3: The relationship between infection with the bacterium Shigella and sex.

Category / sex	adult			children			total		
	total	infected	incidence of infection	total	infected	incidence of infection	total	infected	incidence of infection
males	27	5	12.5%	60	18	12.8%	87	23	25.3%
females	13	2	5%	80	5	3.5%	93	7	8.5%
Total	40	7	17.3%	140	23	16.3%	180	30	33.8%

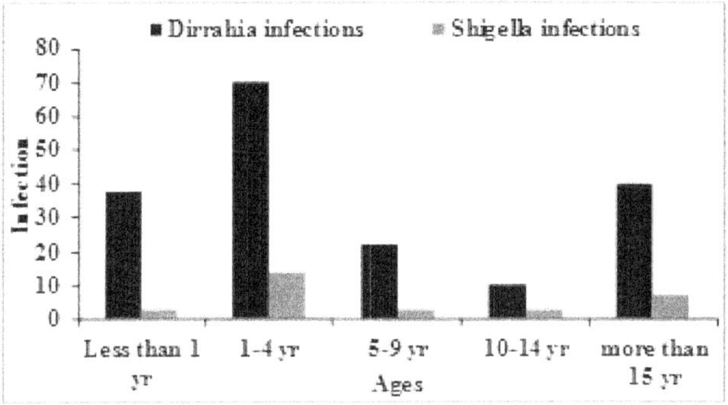

Fig. 1: The relationship between infection with the bacterium Shigella, age and stage.

AGE DISTRBUTION OF PATEINTS SURVYED

During the study 30 bacterial isolates were obtained belonging to the genus Shigella. The total samples amounting to 180 samples were taken. 23 of them were isolated from children less than five years of age. Only seven isolates were taken from adult patients.

Most of the subjects that were infected by Shigella were between one to four years of age. The number of isolates in this category was fourteen bacterial isolates. This was consistent with [1, 3, 5, 12] where injuries were more concentrated in ages one to five years. Table 2 and Figure (1) show the age distribution of patients infected with the bacterium Shigella. Abu-Elyazeed et. al, 2014 [5], has also noted that the lowest rate of infection with the bacterium Shigella occurs in ages less than a year. The prevention includes breastfeeding which protects children from injury, as well as lack of exposure to the causes of injury and this is consistent with our results. There are only three isolates of Shigella bacteria that were isolated from children less than a year old.

Table 4: The sensitivity of isolates of Shigella Spp. (R= Resistant, S=Sensitive)

		Amp	Amp	C	Cip	SxT	Rif	TE	Na	CTX	N
Shigella Spp	1	R	R	R	S	R	R	R	S	S	S
Shigella Spp	2	R	R	R	S	R	R	S	S	S	S
Shigella Spp	3	R	R	R	S	R	R	S	S	S	S
Shigella Spp	4	R	R	R	S	R	R	S	S	S	S
Shigella Spp	5	R	R	R	S	R	R	R	S	S	S
Shigella Spp	6	R	R	R	S	R	R	R	S	S	S
Shigella Spp	7	R	R	R	S	R	R	R	S	S	S
Shigella Spp	8	R	R	R	S	R	R	R	S	S	S
Shigella Spp	9	R	R	R	S	R	R	R	S	S	S
Shigella Spp	10	R	R	R	S	R	R	R	S	S	S
Shigella Spp	11	R	R	R	S	R	R	R	S	S	S
Shigella Spp	12	R	R	R	S	R	R	R	S	S	S
Shigella Spp	13	R	R	R	S	R	R	R	S	S	S
Shigella Spp	14	R	R	R	S	R	R	R	S	S	S
Shigella Spp	15	R	R	R	S	R	R	R	S	S	S
Shigella Spp	16	R	R	R	S	R	R	R	S	S	S
Shigella Spp	17	R	R	R	S	R	R	R	S	S	S
Shigella Spp	18	R	R	R	S	R	R	R	S	S	S
Shigella Spp	19	R	R	R	S	R	R	R	S	S	S
Shigella Spp	20	R	R	R	S	R	R	R	S	S	S
Shigella Spp	21	R	R	R	S	R	R	R	S	S	S
Shigella Spp	22	R	R	R	S	R	R	R	S	S	S
Shigella Spp	23	R	R	R	S	R	R	R	S	S	S
Shigella Spp	24	R	R	R	S	R	R	R	S	S	S
Shigella Spp	25	R	R	R	S	R	R	R	S	S	S
Shigella Spp	26	R	R	R	S	R	R	R	S	S	S
Shigella Spp	27	R	R	R	S	R	R	R	S	S	S
Shigella Spp	28	S	S	S	S	R	R	R	S	S	S
Shigella Spp	29	S	S	S	S	R	R	R	S	S	S
Shigella Spp	30	R	R	R	S	R	R	R	S	S	S
Percentage		7%	7%	7%	100 %	0%	0%	10%	100 %	100 %	100 %

SENSTIVITY OF SHIGELLA SPP. TO ANTIBIOTICS

Sensitivity to antibiotics (10) was tested for and results showed that there is a clear contrast between resistance isolates and the antibiotics used. Results shown in table (4) suggest that isolates of the bacterium Shigella showed resistance by (100%) to the antibiotics (Rifampicin, Cotrimoxazol), and were sensitive by (100%) to the antibiotics (Nalidixic acid, Cefataxim, Ciproflaxacin, Neomycin). Different isolates have different resistance and sensitivity to the antibiotics (Tetracyclin, Amoxicillin, Ampicillin, Chloramphincol), and this is consistent with the studies [7, 10, 19, 20] and with the results of [21]. The resistance to antibiotics (Amoxicillin, Sulphamethaxazol, Trimethoprim) is the most among strains belonging to the genus Shigella. Our results also agreed with the findings of the [22]. The reason for increasing the amount of resistance among isolates belonging to the bacteria Shigella in many parts of the world is due to the indiscriminate use of anti-life in the prevention and treatment of diseases. [7, 10, 11, 23].

CONCLUSION

Our study concluded that in the city of AL- Miqdadiyah, the injuries that Shigella spp. caused are common in the ages 1-5 years old.

ACKNOWLEDGEMENT AND FUNDING INFORMATION

I would like to propose my acknowledgment to AL- Miqdadiyah General Hospital and Al Zahra Hospital for their efforts and assisting in the completion of this study.

REFERENCES

[1] David A. Sack, Christine Lyke, Carol McLaughlin and Voravit Suwanvanichkij. (2001) Antimicrobial resistance in shigellosis, cholera and campylobacteriosis Johns Hopkins. University School of Hygiene and Public Health Baltimore, MD, United States of America.

[2] Eugene, W. N.; Denise, G. A.; Roberts, C. E.; Nancy, N. P.; Martha, T. N. (2001). Microbiology a human perspective.3rd edition. McGraw-Hill companies, Inc.

[3] Beletshachew S.; Sue S.; Ruthanne M.; Suzanne S.; Dana S.; Felicia P. H.; and Thomas Van G. (2004) Trends in Population-Based Active Surveillance for Shigellosis and Demographic Variability in Food Net Sites, 1996–1999. Active Surveillance for Shigellosis. Clinical Infectious Diseases. 38 (Suppl 3): S175–80.

[4] Collee, J.G.; Marmion,B.P.; Fraser, A.G. and Simmons, A. (1996). Practical Medical Microbiology. 14th ed. McGraw-Hill companies, Inc.

[5] ABU-Elyazeed, R.R.;Thomas, F.W.; Robert,W.F.; Shannon , D.P.; Malla,R.R.; Stephen, J.S.; Karim ,A.K.; Leonared ,F.P.; JR.;I brahim ,A.A.; Sahar ,A.; Abdollah ,B.N. and John ,D.C.(2004). Epidemiology of Shigella -association diarrhea in rural Egyptian children. Am J.Trop.Med.Hyg., 71(3), Pp: 367-372.

[6] John, M. A.; Faruque, A.S.; Frauque, S.M.; Sack, R.B. and Mahalanabis, D. (1999). Case-control study of enteropathogens associated with childhood diarrhea in Dhaka, Bangladish. J.of Clinical Microbiology Vol.37, No.11.Pp 3458-3464.

[7] Khalil, k; Khan,SR.; Mazhar, K.; Kaijser, B. and Lindblom, GB. (1998). Occurrence and susceptibility to antibiotics of Shigella species in stool of hospitalized children with bloody dirrhea in Pakistan. Amrican Journal of Tropical Medicine and Hygiene, Vol.58, Issue 6, 800-803.

[8] Qi J.; Zhenghong Y.; Jianguo X.; Yu W.; Yan S.; Weichuan L.; Jinhua W.; Hong L.; Jian Y.; Fan Y.; Xiaobing Z.; Jiyu Z.; Guowei Y.; Hongtao W.; Di Q.; Jie D.; Lilian S.; Ying X.; Ailan

Z.; Yishan G.; Junping Z.; Biao K.; Keyue D.; Shuxia C.; Hongsong C.; Zhijian Y.; Bingkun H.; Runsheng C.; Dalong M.; Boqin Q.; Yumei W.; Yunde H. and Jun Y. (2002). Genome sequence of Shigella flexneri 2a: insights into pathogenicity through comparison with genomes of Escherichia coli K12 and O157. Nucleic Acids Research, Vol. 30, No. 20 4432-4441.
[9] Behram, R.E.; Kliegnam, R.M.; and Jenson, H.B. (2000). Niloson Text Book of Pediatrics.16th ed. W.B. Saunders Company. Philadelphia.
[10] Sally, A. T.; Shelley, N. L.; Harry, S.; Kumar, R.; and Ben, A. (2003). Molecular Epidemiology of the SRL Pathogenicity Island. Antimicrobial agents and chemotherapyvol.47, No.2, Pp. 727–734.
[11] Yeong-Sheng L.; Ming-Ching L.; Ching-Fen K.; Cheng-Hsiung L. and Yi-Hsiung T. (2005). Molecular Epidemiology of Shigella flexneri in a Long-Stay Psychiatric Nursing Center during 2001 to 2003. J. of Clin. Micro., Vol. 43, No. 3, Pp. 1353–1360.
[12] Kotloff, K.L; Winickoff, J.P.; Ivanoff, B.; Clemens, J.D.; Swerdlow, D.L.; Sansonetti, P. J.; Adak, G.K. and Levine, M.M. (1999). Global burden of Shigella infections: implications for vaccine development and implementation of control strategies. Bulletin of the World Health Organization, 77 (8) World Health Organization.
[13] Pickering,L.K.; Peter,G.; Baker,C.J.; Geber,M.A.; MacDonald,N.E. and Orenstein,W.A. (2000) Red Book: Report of the Committee on infectious Diseases. 25 ed. American Academy of Pediatrics.
[14] Cruick shank, R.; Duguid, J.P.; Mormion, B.P. and Swain, R. (1975). Medical Microbiology Vol.2, 12th ed. Edinburgh: Churchill Livinggstone.
[15] Baron, E.J. and Finegold, S.M. (1994) Baily and Scott's Diagnostic Microbiology. 8th ed.C.V. Mosby Company ST.Louis Baltimore Philadelphia. Toronto.
[16] Bauer, A.W.; Kirby, W.M.; Sherris, J.C.; and Turch, M. (1966) Antibiotic susceptibility testing by standardized disk method.Aln.Clin.Pathol. 45 Pp: 493-496.

[17] Stoll, B. J.;Glass, R. I.; Huq, M. I.; Khan, M. U.; Banu, H.; and Holt, J. (1982). Epidemiology and Clinical Features of Patients Infected with Shigella Who Attended a Diarrhea Disease Hospital in Bangladesh. J. Infect. Dis. 146 (2) Pp: 177-183.
[18] al-Khafaji, Sinai Mehdi (2002), the effect of some factors virulent bacterium Shigella spp. On Omraditha. Message submitted to the Faculty of Science / Mustansiriya University as part of a master's degree in life science requirements.
[19] Navia,M.M.;Capitano,L.;Ruiz,J.;Vargas,M.;Urassa,H.; and Schellemberg,D. (1999). Typing and characterization of mechanisms of resistance of Shigella spp. Isolated from feces of children less than 5 years age from Ikara Tanzania. J.Clin.Microb.37 (10) Pp: 3113-3117.
[20] Antoinette, B.; Hartman, Idongesit, I.; Essiet, Danie, W.; Isenbarger, and Luther, E. L. (2003), Epidemiology of Tetracycline Resistance Determinants in Shigella spp.and Enteroinvasive Escherichia coli: Characterization and Dissemination of tet (A)-1 J. of Clin. Microb. Vol. 41, No. 3 Pp. 1023–1032.
[21] Torres, M.E.; Pirez, M.C.; Schelotto, F.; Varela, G; Parodi,V.; Allende, F.; Falconi, E.; Dell'Acqua, L.Gaione, P.; Mendez,M.V.; Ferrari,A.M.; Montano,A.; Zanetta, E;Acuna,A.; Chiparelli,H.;and Ingold,E. (2001). Etiology of Children's Diarrhea in Montevideo, Uruguay: Associated pathogens and Unusual Isolates. J.Clin. Micro. 39 (6) Pp: 2134-2139.
[22] Replogle, M.L.; Fleming, D.W.; and Cieslak, P.R. (2000) Emergence of Antimicrobial Resistant of Shigellosis in Oregon.Clin.Infect.Dis.30 (3) Pp: 515- 519.
[23] Shears, P. (2001). Antibiotic resistance in the tropics transactions of the royal society of tropical medicine and hygiene. 95 Pp: 127- 130.

United Scholars
Publications

Email: info@unitedscholars.net
www.unitedscholars.net

Printed in Dunstable, United Kingdom

91343003R00087